WARS,
GUNS,
and
VOTES

WARS, GUNS, and VOTES

Democracy in Dangerous Places

PAUL COLLIER

HARPER ● PERENNIAL

NEW YORK ● LONDON ● TORONTO ● SYDNEY ● NEW DELHI ● AUCKLAND

HARPER ● PERENNIAL

A hardcover edition of this book was published in 2009 by HarperCollins Publishers.

HarperCollins books may be purchased for educational, business, or sales promotional use. For information please write: Special Markets Department, HarperCollins Publishers, 10 East 53rd Street, New York, NY 10022.

FIRST HARPER PERENNIAL EDITION PUBLISHED 2010.

Designed by William Ruoto

The Library of Congress has catalogued the hardcover edition as follows:

Collier, Paul.
Wars, guns, and votes : democracy in dangerous places /
Paul Collier.—1st ed.
p. cm.
ISBN 978-0-06-147963-2
1. Political violence—Developing countries. 2. Power (Social sciences)—Developing countries. 3. Democracy—Developing countries. I. Title.
HN981.V5C65   2009
306.209172'4—dc22   2008022379

ISBN 978-0-06-147964-9 (pbk.)

10 11 12 13 14   OV/RRD   10 9 8 7 6 5 4 3 2

For John Githongo: his struggle

CONTENTS

WARS,
GUNS,
— *and* —
VOTES

Introduction:

DEMOCRACY IN DANGEROUS PLACES

M Y SON DANIEL, NOW AGE seven, may live to see the eradication of war. Or he might die in one. Why each of these is a realistic prospect for today's children is the subject of this book. War, like disease, has been endemic since the dawn of man. Diseases are now being conquered: in 1977 scientific advance and public action in combination eradicated smallpox. For the first time in history, the world economy looks capable of delivering the material conditions necessary for global peace. But global prosperity also increases the risks: an interconnected world is more vulnerable to any remaining pockets of chaotic violence. Just as the eradication of smallpox depended upon harnessing science through public action, so rising prosperity must be harnessed to secure the prize of global peace.

Wars, Guns, and Votes is about power. Why focus on power? Because in the impoverished little countries at the bottom of the world economy that are home to a billion people, the predominant route to power has been violence. Political violence is both a curse in itself and an obstacle to accountable and legitimate government. It is a curse because the process of violent struggle is hugely destructive. It is an obstacle because where power rests on violence, it invites an arrogant assumption that government is there to rule rather than to serve. You only have to look at the official photographs of political leaders to get the point. In the mature democracies our political leaders *smile*: they are desperate to ingratiate themselves with their masters, the voters. In the societies of the bottom billion the leaders do not smile: their official portraits stare down from every public building, every schoolroom, with a menacing *grimace*. They are the masters now that thankfully the colonialists have gone. *Wars, Guns, and Votes* investigates why political violence is endemic in the bottom billion and what can be done to curtail it.

Since the end of the Cold War two extraordinary changes have occurred, each of which may be opportunities for a decisive shift away from political violence. Both were consequences of the fall of the Soviet Union.

Elections spread across the bottom billion. The image of the popular uprisings in Eastern Europe inspired pressure for political change around the developing world. In the early 1990s national conventions sprang up around West Africa. By 1998 Nigeria, Africa's largest society, sprang out of military dictatorship. Just as around the first millennium the leaders of Europe's petty states had suddenly all converted to Christianity to get in step with the times, so around the second millennium the leaders of the petty states of the bottom billion all converted to elections. Prior to the end of the Cold War most leaders of the bottom billion had come to power through violence: success in armed struggle or a coup d'état. Now most are in power through winning elections. Elections are the institutional technology of democracy. They have the potential to make governments both more accountable and more legitimate. Elections should sound the death knell to political violence.

The other encouraging change is an outbreak of peace. For the thirty years prior to the end of the Cold War, violent conflicts were breaking out more rapidly than they were ending, so that there was a gradual proliferation of civil wars. Once started, civil war proved highly persistent: a civil war typically lasted more than ten times as long as an international war. But then, one after another of the ghastly and persistent civil wars came to an end. The war in Southern Sudan was closed by a peace settlement. The war in Burundi was similarly coaxed into a negotiated peace. The war in Sierra Leone was ended by international peacekeepers. The end of the Cold War unblocked the international community to exert itself against the continued struggle for power by means of violence.

The wave of peace settlements reinforced the wave of elections and promised a brave new world: an end to the pursuit of power through violence. How can we tell how these changes will play out? Can we do more than speculate? I think we can. Although the coincidence of these shocks is unprecedented, each can be analyzed based on how they have played out in the past. There have been previous experiences of electoral competition in the bottom billion. There have been many post-conflict situations. This book uses those experiences to analyze history in the making. As you read *Wars, Guns, and Votes* you may be struck by how fast the research frontier is moving. I get that sense morning by morning as I walk to work wondering whether, during the previous evening, Pedro, or Anke, or Dominic, or Lisa, or Benedikt, or Marguerite has cracked whatever problem we had crashed into by the time I left for home. I hope you get a sense of it too.

Political violence is one variant of the struggle for power. We now see it as illegitimate: might does not make right. In the high-income societies over the past century we have internalized the principles of democracy, and gradually we have come to regard them as universal. Ballots, not bullets, should pave the route to power. Since the end of the Cold War the high-income democracies have taken a further step: from merely regarding these standards as universal to actively promoting them. Despite the tensions over Iraq about whether active promotion should go all the way to enforced regime change or stop short at nonviolent encouragement and inducements, the international community is agreed on the goal. And it has largely succeeded: in the brief period of less than two decades democracy has spread across the low-income world. So what have been the consequences for peace?

The good news is that the world has been getting safer. In fact, despite the catastrophic period of the world wars, it has unsteadily but gradually been getting safer ever since humanity started. Contrary to all those images of the noble savage, early societies were murderous. There never was a peaceful Eden from which we have fallen: peace is something that has gradually been built, millennium by millennium, century by century, and decade by decade. The need for security from political violence has always been fundamental to human society. The great archaeological legacies of antiquity, such as the Great Wall of China and the massive barrier constructed across Jutland by the ancient Jutes against the Germanic tribes, stand as an enduring testimony to the overwhelming priority afforded to collective defense. This priority continued until very recently: for forty years the richest society on earth, America, devoted up to 9 percent of its national income to defense spending to meet the security threat from the Soviet Union.

With the collapse of the Soviet Union an era is over. Despite appearances, the last decade has been rather peaceful. The measure used in this grim academic niche is *battle-related deaths*. The Armed Conflict Data Set keeps a running tally both of the really large conflicts, those that cause at least a thousand such deaths during a year, and of the smaller ones that nevertheless caused more than twenty-five deaths. Here is what happened according to these measures.

Back during the time of late colonialism—1946 to 1959—the number of wars was running at around four a year and the minor conflicts at around eleven. From decolonization to the end of the Cold War in 1991 there was a pretty remorseless escalation. By 1991 there were an astonishing seventeen wars and thirty-five minor conflicts in various parts of the world running at the same time. If violence had continued to spread at that rate, by now we would be facing a nightmare. Instead, 1991 turned out to be a peak. The world is not as peaceful as during late colonialism but we are down to five ongoing wars and twenty-seven minor conflicts. So this break in trend looks to be consistent with the triumph of democracy: where people have recourse to the ballot they do not resort to the gun.

I have come to regard this comforting belief as an illusion. Our approach to political violence has been based on the denial of reality. In consequence there is a brave new world of electoral competition in ethnically divided societies, some of which have just emerged from years of civil war. From 1991 onward the visible trappings of democracy became increasingly fashionable. A president who had not been elected began to look and presumably to feel like the odd one out. It went beyond fashion: many donors began to skew their aid away from unelected governments. And so incumbent presidents braced themselves and decided to face the voters, sometimes emboldened by the knowledge that their people loved them. Sometimes the voters did not do the decent thing.

In the face of voter ingratitude presidents gradually learned how to adapt to the new circumstances. One or two got caught out before they could win. The first was the decent autocrat Kenneth Kaunda of Zambia, who staged an election and lost resoundingly in 1991. At the time of writing, the most recent elections in a society of the bottom billion were those in Kenya, in December 2007. Shortly there will be an election in Zimbabwe. In the years following the defeat of Kaunda, incumbent presidents learned how to win. The Kenyan elections were won by the incumbent, President Kibaki. But within Kenya this was not hailed as a triumph of democracy. Koki Muli, the head of Kenya's Institute for Education in Democracy, had offered the following description: "It is a coup d'état."* As for the elections in Zimbabwe, you have the advantage over me since you know the

---

* "Kabaki Win Spurs Kenya Turmoil," *Financial Times*, December 31, 2007, p. 6.

result. I had no idea who would win the American election of 2008, but I had a pretty clear idea about the outcome of the Zimbabwean elections: I confidently expected that President Mugabe would be reelected. Presidents have discovered a whole armory of technology that enables them to retain power despite the need to hold elections. These elections play out in the context of weak checks and balances, ethnic divisions, and post-conflict tensions.

The triumph of the post–Cold War international community, settlements of the accumulated civil wars of the post-colonial era, is at the same time an alarming point of fragility. Post-conflict situations are dangerous. Historically, many of them have reverted to violence within the first decade. Increasingly since the 1990s, the healing balm for post-conflict tensions and hatreds upon which the international community has relied, and indeed insisted, has been an election. After all, an election should confer legitimacy upon the victor, and the need to secure votes should ensure that the victor has reached out to be inclusive. That comforting strategy has been based upon the denial of an increasingly evident reality.

If the problem of political violence is going to be addressed, we have to understand why small and impoverished countries are so dangerous. To face the reality of political violence we need to understand its technologies: guns, wars, and coups. I know that guns don't kill people: people kill people. A government can conduct a very effective pogrom without any guns at all. The slaughter in Rwanda was done with machetes. But in a violent struggle between organized groups, the one with more guns will tend to win: guns do make violence a whole lot easier. And so I start with guns: both their supply and their demand turn out to be bizarre stories. There is an illicit trade in Kalashnikovs that furnishes supplies, and arms races in Lilliput that drive demand.

War has not yet passed into history, but it now happens "elsewhere." Rich countries no longer fight each other, and they no longer fight themselves. Among the middle-income countries war has virtually disappeared. Even the big poor countries are now pretty safe: China and India have massive armies, but they haven't used them against each other for more than forty years. The world may not hold the line on nuclear proliferation: from time to time more middle-size powers may wish to posture on the world stage by acquiring nuclear capabilities. But over the past sixty years the *first use* of nuclear weapons has built up into a formidable taboo that I cannot see any state breaking.

With the arrival of peace among the more powerful countries, the scale of warfare has diminished: we now have small wars in small countries. Usually the violence is internal: the country tears itself apart while the rest of the world watches. Sometimes the violence draws others in, mostly the neighbors, and sometimes the local regional power. Occasionally the international powers intervene: to prevent internal mayhem, as in the Democratic Republic of the Congo; to expel an invader, as in Iraq 1; or to force regime change, as in Iraq 2. The uncomfortable fact is that a large group of impoverished little countries remain *structurally* dangerous. Wars in the bottom billion are nasty, brutish, and long. They are civil wars; their victims are mostly civilians and they last more than ten times as long as international wars. Although the incidence of civil war has dropped, this is because of a wave of peace settlements: there is still the same momentum for new conflicts to start. Quite aside from the conflicts that were not settled, in 2004 four new wars started up. The following year looked a little better, just one new war. But this was not a peaceful year: there were eight new minor conflicts. Wars were back in business in 2006 with three new ones.

Political violence does not have to take the form of warfare with its attendant "battle-related deaths" to achieve its goal of attaining power. Indeed, the most common and effective form of political violence often succeeds without any deaths at all: it is the surgical strike in the form of a coup d'état. The military, whose purpose is to defend citizens from organized violence, is sometimes in a splendid position to perpetrate it. Globally since 1945 there have been some 357 successful military coups. And for each successful coup there are a lot of failures. For Africa, the one region for which there is a comprehensive tally, in addition to the 82 successful coups there were 109 attempted coups that failed and 145 coup plots that got nipped in the bud before they could even be attempted. That is around seven planned surgical strikes for the average country. In many societies presidents are more likely to lose power to their army than through any other route.

Guns, wars, and coups have been the reality of the bottom billion. They have destroyed societies that were confidently expected to develop. The meltdown of Cote d'Ivoire, once the most celebrated society in Africa, shows all three of these technologies in ruinous action over the course of a decade.

Does it matter if political violence in its various manifestations continues to be the predominant route to power? Perhaps the whole notion of exporting our democratic values to these societies was merely a comfortable delusion and they are better left as they were? Of course it matters.

For one thing our democratic values *are* universal. Governments are not there to command their citizens: they are there to serve them. The journey from citizen servitude to government servitude has been a long one in our own societies. It will probably be a long one in the societies of the bottom billion. We have most surely underestimated the degree of difficulty and promoted the wrong features of democracy: the façade rather than the essential infrastructure. I will argue that in situations in which it is not feasible to build the infrastructure, creating the façade is likely to frustrate democratic accountability rather than fast-track it.

It matters because in the divided societies of the bottom billion, when political power is won through violence, the results are usually awful. The political strongman in a divided society is seldom a visionary leader; he is more likely to be self-serving, or in thrall to the interests of a narrow support group. Visionary leadersh͏ portant, but its role is to turn states into nations. The fun͏ mistake of our approach to state building has been to forget that well-functioning states are built not just on shared interests but on shared identity. Shared identity does not grow out of the soil; it is politically constructed. It is the task of political leadership to forge it.

It matters because the process of violent struggle for power is hugely costly. Wars and coups are not tea parties: they are development in reverse. Wars may now be small in the sense of few "battle-related deaths," but the increasing involvement of civilians, and indeed the blurring of the distinction between civilians and combatants, implies that even small wars can have highly adverse consequences. Political violence is not just a curse for the societies in which it occurs; it is an international public bad. Most particularly, it damages the neighbors, something that has profound implications for sovereignty.

The overarching problem of the bottom billion is that the typical society is at the same time both too large and too small. It is too large in the sense that it is too diverse for cooperation to produce public goods. It is too small in the sense that it cannot reap the scale economies of the key public good, security. But the only point of understanding the nature of the problems is that it helps in the search for effective solutions. If the problem is that societies are too large to have an inherited sense of common identity, state building is not, fundamentally, about institutions, which is the fashionable nostrum. There is a prior essential stage of nation building that takes more visionary leadership than has been forthcoming in most of these societies.

If the problem is that societies are too small to supply key public goods, then it is pointless to place national sovereignty on a pedestal. Given the structural deficiencies in their states, the citizens of the bottom billion have little choice but to have recourse to the international supply of essential public goods. To some extent they can do this by pooling their sovereignty, something that to date they have singularly failed to do. But that failure is itself symptomatic: much of the supply of the international public goods that the bottom billion need is going to have to come from the countries that already know how to cooperate to supply such goods: the high-income countries. Yet the indignant defense of sovereignty by the governments of the bottom billion, combined with the pusillanimity and indifference of leaders in high-income countries, radically constrains what international action can realistically achieve. The core proposal of this book is a strategy whereby a small intervention from the international community can harness the political violence internal to the societies of the bottom billion. This powerful force that to date has been so destructive can be turned to advantage, becoming the defender of democracy rather than its antithesis.

To harness the political violence inherent in the societies of the bottom billion as a force for good, we will need a very limited use of international force. After Iraq, international peacekeeping provided by the forces of the high-income countries is unpopular, both with voters in the high-income world and with alarmed governments of the bottom billion. But military intervention, properly constrained, has an essential role, providing both the security and the accountability of government to citizens that are essential for development.

I am aware that I walk a tightrope. Those who regard the societies of the bottom billion as an irredeemable quagmire will be predisposed to regard the proposals in this book as costly idealism. Those who regard these societies as the victims of neo-imperialism will be predisposed to regard the proposals as imperialism in disguise. Above all, those who regard internal political violence in any form as illegitimate will be predisposed to regard the proposal for harnessing it as breaching a fundamental tenet. But the proposals in this book are not costly idealism: they are grounded in analysis and evidence. Nor are they a backdoor form of imperialism. Citizens of the bottom billion have the same rights as the rest of us, including a legitimate aspiration to nationhood. Nor do they undermine the tenets of democracy. My message is that the aspirations to nationality and democracy cannot be achieved by the path currently being taken: fake democracy protected by the sanctity of sovereignty is a cul-de-sac. Just as the high-income world should provide a vaccine against malaria for the citizens of the bottom billion, so it should provide them with security and accountability of government. All three are public goods that will otherwise be chronically undersupplied. Only once they are properly supplied can the societies of the bottom billion achieve their aspirations to genuine sovereignty.

The defeat of political violence is where our illusions are most inextricably bound up with our hopes and our strategies. And it is where our errors, grounded in those illusions, are proving most costly. Each of the changes I analyze is potentially hugely hopeful. But it turns out that each is a two-edged sword. They might well trigger processes that substantially increase violence. But it is not simply a story of "things might go wrong." Within the limits imposed by modern research methods, I think I can show what will determine whether democracy is going to be transformative or destructive. More alarmingly, to date democracy in the societies of the bottom billion has increased political violence instead of reducing it. But my message is not meant to denigrate the efforts of brave people who have struggled for their democratic rights: I am not an apologist for dictatorship. Only by moving on from illusion can we work out what practical measures could harness the undoubted potential of democracy as a force for good.

Part I

DENYING

REALITY:

DEMOCRAZY

Chapter 1

VOTES AND VIOLENCE

OUR TIMES HAVE SEEN A great political sea change: the spread of democracy to the bottom billion. But is it democracy? The bottom billion certainly got elections. They were heavily promoted by American and European pressure, and, as the most visible feature of democracy, they were treated as its defining characteristic. Yet a proper democracy does not merely have competitive elections; it also has rules for the conduct of those elections: cheating gets punished. A proper democracy also has checks and balances that limit the power of a government once elected: it cannot crush the defeated. The great political sea change may superficially have looked like the spread of democracy, but it was actually the spread of elections. If there are no limits on the power of the winner, the election becomes a matter of life and death. If this life-and-death struggle is not itself subject to rules of conduct, the contestants are driven to extremes. The result is not democracy: I think of it as *democrazy*.

The political system that preceded democrazy was personal dictatorship. Usually it did not have even the veneer of an ideology. Personal rule reached its apogee in President Mobutu of Zaire, whose extraordinary system of government is depicted in Michela Wrong's *In the Footsteps of Mr. Kurtz*. Personal rule meant ethnic favoritism and the erosion of the institutions of the state. Mobutu's power came to rest on greed and fear: his patronage might reward loyalty with unseemly wealth, and his thugs might punish suspected opposition with torture. Where there was an ideology it was Marxist, such as the Derg regime in Ethiopia, and the MPLA in Angola; grim and ruinous regimes that attracted a predictable swath of support among the Western left. More commonly the Marxist ideology was a decorative veneer, a language of politeness appropriate for the circles in which political leaders mixed, much as Christian sentiments must have been de rigueur in a nineteenth-century drawing room. In Zimbabwe, where this make-believe blossomed, there was a politburo and everyone was referred to as comrade. Such undemocratic regimes looked as though they were inviting violent opposition. Mobutu and the Derg were both overthrown by rebellions, and the MPLA faced a huge uprising from UNITA.

Across Africa, Latin America, and Asia during the 1990s, autocracies fell like ninepins. Sometimes citizens took heart from the example of Eastern Europe and massed in the streets, the most stunning instance being the overthrow of President Suharto in Indonesia. Sometimes aid donors made further funding conditional upon democracy, the best-established instance being Kenya, where the diplomatic community recognized that President Moi could be pressured. Sometimes autocrats saw which way the wind was blowing and decided to go with the flow. Autocrats commonly surround themselves with sycophants, and this probably helped the process of democratization on its way. Imagine what an autocrat who is contemplating democratization is going to ask his entourage. There is really only one question: if I hold an election, would I win? And what can a sycophant say? Quite possibly the sycophant has no clue: it has not been his job to gauge public opinion. However, even if he suspects that people detest the president, he has a problem. Hasn't he been telling the president for years how much his people love him? Those advisers who told the president the truth tended not to last long as advisers.

At least three autocrats got caught this way, Suharto in East Timor, Kaunda in Zambia, and Mugabe in Zimbabwe. All let citizens vote because they were sure they would win. Suharto lost East Timor as a result: people voted overwhelmingly for independence. Kaunda did a little better than Suharto: he managed to get about 20 percent of the vote, so some people did indeed love him, namely those in his home region, which he had favored with public spending. As the results came in he was naturally outraged that citizens had been so ungrateful. Quite what might have happened at that point we will never know. Fortunately, Jimmy Carter was in the country leading a team of election observers. As the results started to come in, Carter sensed what to do. Rushing to the presidential palace, he felt Kaunda's pain and stayed there until it was too late to annul the election. After all, he had lived through a similar experience. With Carter there in the palace, Kaunda had little choice but to accept the defeat. Whether he would have done so without Carter is an open question: reputedly he then went around the capitals of Africa advising presidents not to make his mistake.

And President Mugabe? By the mid-1990s President Mugabe had followed the fashion, adopting a constitution in which there were multiparty elections and term limits on the presidency. Many dictators agreed to term limits, confident that by the time the limit was due to bind they could change the constitution by one means or another. And so term limits turned into time bombs. President Putin of Russia is, of course, the most spectacular example of a successful constitutional side step: don't even bother to change the term limit, make yourself prime minister and shift effective power from the presidency to the new position. President Obasanjo of Nigeria tried but failed to extend his term, as did President Chiluba of Zambia. Presidents Deby of Chad and Museveni of Uganda were more successful. President Mugabe decided to change the constitution, removing the term limit and drastically increasing presidential powers. To do this he needed a referendum. It was this that he lost.

Unfortunately, the referendum did not coincide with a presidential election, and so Mugabe continued as president, now knowing that he would lose a democratic election. I will return to the problem he faced shortly. For the present I want to stay with the spread of democracy. Country by country, governments subjected themselves to competitive elections. Sometimes they won, sometimes they lost, but either way, opposition was now better able to express itself.

So how has this spread of democracy affected proneness to political violence? Pretty obviously violence should go down. It may be obvious, but in general it helps to spell out the basis for what we think we know. There seem to me to be two reasons for expecting democracy to reduce the incidence of political violence. I will call them accountability and legitimacy, and they are complementary and so reinforcing. The accountability effect works as follows. In a democracy a government has no choice but to try to deliver what ordinary citizens want. If it is seen to perform sufficiently well, then it gets reelected; if it is judged to be inferior to alternatives, then it loses. Either way, government strives to perform because it is accountable to voters. A dictator might choose to deliver performance that is just as good as this, but for the dictator it is just that, a choice. The democratic government has no option. And in practice, all too often dictators choose to do something completely different, as with Mobutu. So democracy tends to improve government performance by subjecting leaders to the discipline of being accountable. Why might this in turn reduce political violence? Well, obviously, because there is less basis for grievance. If the government performs better for ordinary people, then they are less likely to take up arms against it.

So much for the accountability effect, how about legitimacy? Being elected is now widely seen as the only basis for government legitimacy. In turn, at least according to democratic theory, a legitimate government thereby acquires certain rights. A legitimate government has a *mandate* to do what it said it would do, and this entitles it to face down opposition to the implementation of its program, at least within limits. In a democracy citizens agree to these rules, and so opposition to a government's elected program cannot legitimately extend to the use of violence. This provides a further reason for the reduction in political violence. Even if the most extreme opponents of the government do not accept that the government is entitled to enact its program, they will find it more difficult to enlist mass support for violent opposition. They can no longer reasonably claim that their struggle is just.

Democracy should thus deliver a double whammy against political violence: there is less objective basis for grievance, and for any given grievance it should be harder to persuade people to resort to violence against the government.

So confident have we been in asserting that democracy is the answer to political violence that it seems almost churlish to look at the evidence to test whether it is right. The peace-promoting benefits of democracy have become one of the fundamental certainties of the policy world, indeed perhaps one of the few unifying beliefs across the political spectrum. George Soros and George Bush have not agreed on much, but I suspect that they would be on the same side on this one, along with millions of other people.

When the countries of the bottom billion started to democratize I was as enthused as anyone. However, the ensuing years have been more difficult than I had expected. I have little time for outside commentators who turn into tut-tutting judges. Change is difficult and there are strong forces resisting it. It is not that the societies of the bottom billion have failed to live up to my expectations. Rather, I was coming to suspect that I had missed things that in retrospect were becoming evident. Indeed, there had surely been people with doubts all along, but their voices had been drowned out in the cacophony of enthusiasm for democracy. Essentially I came to suspect that theories that were entirely appropriate for countries that were more developed might have been overextended. The societies of the bottom billion may simply be lacking the preconditions whereby the accountability and legitimacy effects were going to work very well. I have to say that I came to these doubts with deep reluctance. But it was time to turn to the evidence.

You might expect that the relationship between democracy and political violence would be settled academic territory. But somewhat to my surprise I found that it was not. It was, in fact, about as close to terra incognita as modern social science gets: I could not find a single published paper. I teamed up with Dominic Rohner, a young Swiss researcher, and got to work.

We got data on virtually all the countries in the world for the period since 1960. Controlling for the other characteristics that were likely to matter, how did democracy affect the incidence of political violence? At first we could find no relationship. To me this nonresult seemed intrinsically unlikely: surely something as salient as the political regime simply had to matter. Then it occurred to us that the relationship might well not be the same across the entire range of economic development. After all, the societies of the bottom billion were highly distinctive in being far poorer than the other democracies. Maybe in poor countries the effect of democracy on violence was not the same as in rich countries. Once we introduced this possibility we found that the political regime always mattered. In fact, democracy had the opposite effect in poor countries to that in rich countries. It was because the two effects were opposing that there had appeared to be no effect at all. So what were the two opposing effects?

We found that in countries that were at least at middle-income levels, democracy systematically reduced the risk of political violence. The prediction of the accountability-and-legitimacy view of how democracy should make a society more tranquil was borne out. But in low-income countries, democracy made the society more dangerous. As if poverty was not miserable enough in itself, the effect of democracy adds insult to this injury. Whereas in societies that are not poor it enhances their already safer conditions, in poor societies democracy amplifies the already severe dangers.

If democracy makes poor societies more dangerous, but societies that are not poor safer, there must be some threshold level of income at which there is no net effect. The threshold is around $2,700 per capita per year, or around $7 per person per day. The societies of the bottom billion are all below this threshold: most of them are a long way below it.

To my mind the key implication of these results was that the accountability-and-legitimacy theory of how democracy would help the societies of the bottom billion must be missing something. Indeed, it must be missing an elephant. Much of this book is devoted to flushing out that elephant. But I have not quite finished with the results of our investigation.

Recall that at higher levels of income societies are safer. It turns out that all the benign effect of higher income depends upon the society being democratic. Indeed, it is more striking than that: in the absence of democracy, as a society starts to get rich it becomes more prone to political violence. Democracies get safer as income rises, whereas autocracies get more dangerous. If it helps, you can think of this as two lines, an upward-sloping one showing how democracies get safer as income rises, and a downward-sloping one showing how autocracies get less safe. The level of income at which democracy has no net effect on violence, $2,700, is simply the point at which these two lines cross over. Applying this to the society with the most astounding income change of our times, China has now passed the income threshold—per capita income has soared past $3,000. So, if China runs to form, year by year its spectacular economic growth is now making it more prone to political violence unless it democratizes.

Our initial work had been pretty heroic in the sense that we had hastened over a host of statistical cans of worms. Much of our work now turned to opening these cans and seeing if the results survived. For example, income is likely to be affected by both conflict and the political regime. Causality might in fact be running in the opposite direction to our interpretation. We checked on this and satisfied ourselves that this was not the explanation: our results were not spurious, at least not on this count. In the small world of the statistical study of political violence, the foremost rival team has been James Fearon and David Laitin at Stanford. Like us, they had a model of the factors that tend to produce violence, but it differed in detail from our own. We decided that a good test of the result that democracy increased the risk of violence for the bottom billion would be to see whether it survived if we introduced it into their model. Unfortunately for these societies, it did survive. To my mind the most remarkable result came when we investigated a range of different forms of political violence. We looked at assassinations, riots, political strikes, and incidents of guerrilla activity as well as full-blooded civil war. To my amazement, the same pattern was true for them all: at low income, democracy increased political violence.

I do not believe that these results reveal unalterable relationships: later I will argue that democracy can be made to work in the societies of the bottom billion. But consider for a moment what would be the implication if they were unalterable. They would imply that judged by the objective of peace, there would be a preferred sequence for economic and political change. The ideal stage at which to democratize would be once a society had already reached a moderate level of development.

As Dominic and I digested these results we started to puzzle over the obvious question: why? The question actually decomposes into three distinct puzzles. First, why was the benign effect of democracy that reduced the risk of political violence dependent upon the level of income: what was it about income that made democracy differentially peace-promoting in richer societies? The second was the converse question of why autocracies become more dangerous at higher levels of income. Finally, and most mysteriously, once these income-related effects of democracy and autocracy were allowed for, there remained a further pure effect of democracy that was making societies more at risk of violence. Like some unobservable dark matter it was lurking as a constant across societies. What was it? These were not easy questions.

The key insight came by the simple psychological technique of imagining myself in the position of being a former dictator in one of the countries of the bottom billion who had caved in to pressure from donors to democratize. How had I kept the peace before and how did democratization change my problem? I was evidently not the first person to wonder about how a dictator might best stay in power. Herodotus reports that when Periander became the young dictator of Corinth, he sent a messenger to the old and experienced dictator of Miletus, Thrasybulus, for advice. Thrasybulus had clung to power very effectively; had he any tips for someone just starting out on the same career? Thrasybulus took Periander's messenger into a field of corn and, as he talked, repeatedly and systematically snapped off the heads of all the tallest stems. The messenger returned baffled, but Periander got it. Although social science has advanced in the two and a half thousand years since Herodotus, I think that this still gives a pretty fair take on the technology of power retention. If we are to generalize from Thrasybulus, the key is to be preemptive: purge potentially dangerous people before they act. Does democracy affect my ability to undertake such purges? Well, the awkward problem with preemptive purges is that they are not compatible with the rule of law: the technique depends upon punishing people even though they haven't done anything. This sort of conduct collides with even fairly modest levels of democracy.

The idea that the ability to mount a purge would be reduced as a result of democracy was a plausible explanation for the dark matter. If leaders could no longer mount preemptive purges they might be less able to keep the lid on political violence. This might be why, over and above those effects of democracy that depended upon the level of income, there was the pure effect that increased political violence. Herodotus had given us an idea; now it was time to test it.

We turned to a large political science data set on purges. Believe it or not, these things are measured, country by country, and year by year. We wanted to see whether democracy made purges more difficult, controlling for other possible influences. Sure enough, even a modest degree of democracy radically reduces the frequency of purges. From the perspective of keeping the peace through repression, democracy is a massive technological leap backward.

If you want a practical, real-world, up-to-the-minute example of how democratization can make it harder to keep the peace, try Iraq. Whatever the limitations of the present regime, it is clearly massively more democratic than that of Saddam Hussein. Yet Hussein presided over a relatively peaceful country. It was not an attractive peace, but it was a peace of sorts, and it most surely depended upon preemptive repression rather than citizen consent.

So a weakening of technologies of repression is, I think, a likely explanation for the dark matter: the higher risk of political violence that comes from democracy. Why, then, should the net effect of democracy be increasingly favorable as income rises? I think the answer lies in those effects that I started with: accountability and legitimacy.

The stark and straightforward reason that in the bottom billion the accountability and legitimacy effects of democracy do not reduce the risk of political violence is that in these societies, democracy does not deliver either accountability or legitimacy. So why does it fail to do so?

OVER THE YEARS I HAVE had some very smart students, but undoubtedly the smartest was Tim Besley, now a highly distinguished professor at the London School of Economics and a former editor of the *American Economic Review*. Tim's book *Principled Agents?* is the most serious theoretical attempt to answer the question of whether having to face voters actually disciplines politicians. It is a complicated book, but I think I can give you the gist of it. In our own societies the answer to Tim's question seems pretty obvious. If an incumbent politician had not even *tried* to deliver what people want, electors would notice. The actions of political leaders are scrutinized by the media, and if a politician were consistently to advance his own interests at the expense of ordinary citizens he would not be reelected. Politicians want to stay in power. Partly, let us hope, this is because they feel a sense of vocation to do good, but also pretty obviously because it is their choice of lifestyle: it is their profession, and they do not want to be unemployed. And so, between media scrutiny and politicians' appetite for power, political leaders are pinioned to trying hard for the common good.

But in the societies of the bottom billion conditions are often not like this at all. Suppose that voters have precious little knowledge about the choices they face. Even the past performance of the incumbent, which voters have just lived through, will typically be open to multiple interpretations. Perhaps bad outcomes were due to mitigating circumstances; perhaps the government was not to blame. All too often, in the volatile economies of the bottom billion, this is genuinely the case: the economy frequently gets derailed by shocks beyond local control. A typical shock is that the price of the country's export good crashes and the economy consequently collapses. I can think of three African democracies in which this happened in the run-up to an election. In each case the incumbent government had done a pretty good job. One was in Benin during the run-up to the 1996 election, removing a reforming president. It happened again in Uganda in the run-up to the 1998 election: the world price of coffee crashed. And it happened in Madagascar before the 2006 election with a combination of falling export prices and soaring costs of imported oil. How was the electorate to tell whether the economy crashing around their ears was crashing because of an unavoidable external shock or because the government had been incompetent? Of course, the government tried to explain, but governments had always made excuses. How were they to know what to believe?

In addition to the problem of lousy information, perhaps some voters are going to vote for or against the incumbent regardless of performance because of their ethnic identity. Identity is the basis of most voting in the bottom billion. Their societies are usually divided into competing ethnic identities, and as a result ethnicity is by far the easiest basis on which to organize political loyalty. The problem with it is that because the loyalty isn't issues-based, it isn't performance-based either. Votes are simply frozen in blocs of rival identities. A consequence of having great blocs of votes frozen into support or opposition is that the vote that an incumbent politician attracts is not very sensitive to performance: few votes hinge on whether he has done a good or a bad job. So not only do people lack the information on which to judge performance, but relatively few are going to base their votes on this judgment.

Perhaps also, the scope for the government to produce a good performance is really quite modest, maybe due to its own limitations. Especially after years of poor performance, a government may simply lose faith in its own ability to make a decisive difference to economic events.

Finally, suppose that if the government does choose to be good it has to forgo behavior that is decidedly lucrative. Messing about with the economy may be detrimental to ordinary citizens, but it opens up many little niches and crannies for personal enrichment, and for rewarding loyalty among followers. If all these opportunities are closed off, the leader has no means of maintaining loyalty.

So how does this stack up? As the quality of voter information is made weaker, as identity politics freezes more and more votes, as the government's confidence in its own ability to shape events diminishes, and as the costs of forgoing bad governance are increased, a point is reached at which facing an election simply does not discipline an incumbent politician into trying to perform well. And if politicians can still face a reasonable chance of winning without bothering to deliver good performance, then—and this is Tim's killer point—the sort of people who seek to become politicians will change. If being honest and competent does not give you an electoral advantage, then the honest and competent will be discouraged. Crooks will replace the honest as candidates.

One depressing indicator of such a process is that democratic politics in the countries of the bottom billion tends to attract candidates with criminal records. You might reasonably expect that having a criminal record would make running in an election a nonstarter. I think it would in America or Britain, and indeed across most of the rich world. But in the societies of the bottom billion it simply isn't so. Electors just don't have enough information to sort out the accusations from reality: either the press is muzzled or it is too free—there is so much mud being slung without recourse to verification that voters discount whatever they are told. Or electors are frozen in ethnic loyalties and so support their own politicians even if they are criminals.

Evidently, one reason elected office is more attractive to criminals than to the honest is that only the criminals will take advantage of the opportunities for corruption. But there is sometimes a further reason: elected office provides immunity from prosecution. Ask yourself for whom this is particularly valuable. For the honest, it merely protects from mischievous attacks that, in the end, they could probably resist anyway. But for the criminal, immunity from prosecution is likely to mean the difference between freedom and jail. Sometimes this turns to farce. Following the Nigerian gubernatorial elections of 2007 there was a race between the police and a victorious deputy governor as to whether he could get himself sworn in before they could reach him to arrest him. It was touch and go whether home would be jail or the deputy governor's villa.

If honest people realize that they are unlikely to win and so do not come forward as candidates, then voters lack even the choice of a decent leader. There is really not much point in finding out about the candidates, and this adds a further twist to the vicious circle.

Tim's analysis is about at the frontier of serious work on democracy. But even Tim's world is thoroughly sedate when compared to the election campaigns familiar to the bottom billion. Basically, in Tim's world politicians still play by the rules; it is just that they face badly informed electors. Again, I put myself in the situation of an old autocrat now having to retain power in a democracy. What options do I face? Hard as it is to bear, I have to be honest with myself that my people do not love me. Far from being grateful for the wonders that I have achieved, they may increasingly be aware that under my long rule our country has stagnated, whereas elsewhere initially similar countries have transformed themselves. There are even a few cogent voices out there explaining why this is my fault. I shake my head in disbelief that it has come to this, seize my gold pen, and start listing the options. I decide to be systematic, in each case putting down the pros and cons.

OPTION 1: TURN OVER A NEW LEAF AND BECOME A GOOD GOVERNMENT

Pros: This is probably what most people want. It would make a change, I might start feeling better about myself, and I might even leave a legacy that my children could be proud of.

Cons: I haven't much of an idea how to do it. The skills I have developed over the years are quite different, essentially how to retain power through shuffling a huge number of people around a patronage trough. My God, I might have to read those damned donor reports. And even if I worked out what needed to change, the civil service isn't up to implementing it. After all,

I've spent years making sure that anyone who was exceptional or even honest was squeezed out: honest people cannot easily be controlled. Yes, I too read Herodotus. Even worse, reform might be dangerous. My friends, the parasitic sycophants with whom I have surrounded myself, might not put up with it: they might decide to replace me in a palace coup. They would probably dress it up to the outside world as reform! But suppose I did it; suppose I actually delivered good government. Would I get reelected? I start to think through all those rich-country political leaders who, over the years, have met me, often lecturing me on the need for good governance. What became of them? What was their record of electoral success? I do a rough tally—they seemed to win their own elections only around 45 percent of the time. So, if I pull it off, I have a 45 percent chance of winning.

Option 1 does not seem that attractive, whatever the foreign ambassadors might imply with their incessant homilies about good governance. The evident difficulties of governing well make your electoral task daunting relative to that of your fortunate rich-country counterparts. You contemplate having a comforting sulk about the inequities of life, but put self-indulgence behind you: you have to make the best of what you have. And then it strikes you that compared with your rich-country counterparts you have one potential advantage. Although you are going to have to win an election, you are not subject to much effective scrutiny as to how you go about it. Does this open up any strategies that might enable you to win despite continuing to be a bad government?

OPTION 2: LIE TO ELECTORS

Pros: You control most of the media, so it is relatively easy. What is more, your citizens have neither education nor good reference points by which to tell how bad things really are. So you can tell them how fortunate they are to have you as president.

Cons: You have been doing this for years and so people heavily discount anything you say.

On balance, although lying seems to be worth doing, you simply cannot rely on it to deliver victory.

OPTION 3: SCAPEGOAT A MINORITY

Pros: This one works! You can blame either minorities within your country or foreign governments for all your problems: that President Mugabe of Zimbabwe is a role model. The politics of hatred has a long and electorally pretty successful pedigree. Most of the societies of the bottom billion have unpopular ethnic minorities to pillory, and failing all else you can always blame America. You can also promise favoritism for your own group.

Cons: Some of your best friends are from ethnic minorities. In fact, they have been funding you for years in return for favors. You prefer business people from ethnic minorities because however rich they become, they cannot challenge you politically. It is the core ethnic groups that you want to keep out of business. If you scare the minorities too badly they will move their money out.

So, although scapegoating works, beyond a certain point it gets rather costly.

OPTION 4: BRIBERY

Pros: Bribery plays to one of your key advantages over the opposition—you have more money.

Cons: Can you trust people to honor the deal? If you pay them money will they actually vote for you? After all, there are some pretty unscrupulous people out there.

On balance you are not sure. If only there was some reliable research evidence! You search the Net and stumble on something by someone called Pedro Vicente, of the Centre for the Study of African Economies at Oxford. You start to skim it and rapidly become riveted, as well you might. Pedro has conducted a randomized, controlled experiment on electoral bribery in São Tomé and Principe, which is just off the coast from your own state.

Tiresomely, you find that the main thrust of his research is to investigate whether bribery can be countered. Then, however, you find the pertinent gem. In some districts bribery was restrained by external scrutiny, whereas in others it was not. Systematically, the candidate who was bribing gathered more votes in those districts where bribery was not restrained. Bribery works!

In fact, bribery comes in two modes: retail and wholesale. Retail bribery is expensive and difficult but may still be worthwhile. Its advantage is that you can target pockets of voters who are critical for success. For example, President Moi of Kenya managed by astute attention to key votes to win an election with only 37 percent support. Why doesn't bribery backfire? If the British Labour Party was caught offering money to individual voters in exchange for their support the electoral damage would be massive. But in many societies elections are viewed differently. Politicians deliver nothing during their period in office, and so people expect that during the one brief moment when they exert some power politicians should

dispense patronage, and hard cash in the pocket is better than promises. But even if politicians can offer bribes without provoking criticism, how can they enforce the deal? After all, the vote is secret.
What is to stop voters from accepting the money and then voting
for the opposition?

In Kenya the opposition recognized that telling people that taking bribes was wrong would be a vote loser and so did not even
attempt it. Instead they proposed that people should take the bribe
from the government but vote for the opposition. Why is such an
opposition message not a very effective counter? The government
has two points of discipline. One, paradoxically, is morality: often,
ordinary decent people feel bad if they take someone's money but
then renege on their undertaking. The opposition argument that
one wrong neutralizes the other is smart, but it is morally a little
tortured. The other is fear of detection: how secret is the ballot? In
Zimbabwe President Mugabe's street boys spread the word that the
government would know how votes were cast, and in the prevailing conditions of misgovernance this could not be treated as an idle
threat.

It is not as if one individual vote will determine the choice of
government: realistically, it will have no effect whatsoever on the
outcome. And so even if there is only a small risk that a vote against
the government may be detected, it may not be worth taking. It
might land the voter in trouble and so be irresponsible for an adult
struggling to bring up a family in conditions that are already dire.

Having got this far in his train of thought, the president will
perhaps be counting his fortune. How much does it cost to bribe
the typical voter, how many votes does he need to buy, and how
much can he afford? In some societies he will sit back contentedly:
this strategy is within his budget. In others he may be pondering
whether there is a cheaper way of buying votes. There is: it is time
for wholesale bribery.

Wholesale bribery works by paying for votes delivered in blocs rather than individually. Bloc voting is very common in impover-ished traditional rural societies: the local big shot gives the lead and his advice is not seriously questioned. When votes are counted it is common for many villages to have voted 100 percent for one candi-date. If the big shot determines the voting, it is obviously cheaper to buy his support directly rather than try to attract individual votes.

Overall, you conclude that bribery is your kind of strategy. The only problem is whether you have enough money to win with it. This inspires you to carry on thinking.

OPTION 5: INTIMIDATION

Most politicians try to ingratiate themselves with voters, but a radically different technique is to intimidate them.

Pros: Most people are not particularly brave, and when con-fronted by thugs threatening personal violence, they back down rather than stand up for themselves. One big advantage of intimidation is that even if you cannot observe *how* people vote, you can observe *whether* they vote. Given that you are playing in identity politics, you know perfectly well the iden-tity of those who intend to vote for your opponent. So you can threaten them that if they vote they will suffer. Does it work? In Kenya President Moi used it to force a mass of Kikuyu living in the Rift Valley who were likely to vote against him to move. In moving they went to areas where they were not registered to vote, so he no longer had to worry about them. He claimed the violence was just a local dispute about land rights, but a careful statistical study by two Kenyan researchers, Mwangi Kimenyi and Njuguna Ndung'u, gave the lie to that one. They show that "the central rationale of the violence appears to have been to maintain the political and economic status quo in the region during the run-up to the general elections."[*] Indeed, the bows and arrows ostensibly used by irate and untamed tribesmen turned out to have been manufactured in East Asia and presumably planted by the government. You also recall that President Mugabe has not been reticent in using intimidation against opposition voters.

Cons: If politics turns violent there is no knowing where it might stop. The other side might turn violent. After all, the other side has the advantage of numbers: if they didn't, you would not have to worry about winning the election. You don't want to risk losing a contest in violence.

Overall, violence might turn out to be a can of worms. The opposition might be even more violent than you are. This is not reason for not doing it: you may well need to do it simply to counter the violence that is coming from the opposition, who are, after all, making the same calculation. But violence may not be enough to ensure that you win.

Option 6: Restrict the field to exclude the strongest candidates

Pros: This is particularly appealing because not only do you increase your chances of winning but you hit directly at the people you most hate: your personal opponents. You have to find some reason for excluding them, but that is not particularly difficult. You can accuse them of corruption—after all, it is quite likely to be true. A delicious nuance is that since the donors are always

---

[*] Mwangi Kimenyi and Njuguna Ndung'u, "Sporadic Ethnic Violence: Why Has Kenya Not Experienced a Full-Blown Civil War?" in *Understanding Civil War (Volume 1: Africa)*, ed. Paul Collier and Nicholas Sambanis (Washington, D.C.: World Bank, 2005).

urging you to be tougher on corruption, they can scarcely object to this option. Even the challengers to those international role models Presidents Obasanjo in Nigeria and Mbeke in South Africa were prosecuted. Admittedly, those prosecutions were probably warranted, but you can still claim to be following their precedents. If corruption is too sensitive an issue to open, you can try citizenship. Given the considerable ethnic diversity of most countries of the bottom billion, and the large migrations of peoples, it should be easy to trump up some ancestry that debars them from citizenship. Potentially, you can go the whole hog, like President Abacha of Nigeria, and debar everyone. Implausible as it might seem, it is still possible to hold a contested election. Failing all else, someone might assassinate your opponent, as happened in the run-up to the Pakistani elections of 2007, which Benazir Bhutto might otherwise have won.

Cons: Unless you go the whole hog, voters inevitably have some alternative to your own good self, however awful. They may be sufficiently foolish to opt for it. You think mournfully of President Gueï of Cote d'Ivoire, whose sad story must wait a little.

So banning key opponents makes sense, but cannot be relied upon to be sufficient. Worried, you wonder whether there is any strategy that you have overlooked. And then you heave a long, deep sigh of relief.

OPTION 7: MISCOUNT THE VOTES

Pros: At last you have found a strategy that sounds reliable. With this one you literally cannot lose: incumbent one, opponent ten million; headline: "Incumbent Wins Narrowly by One Vote." It also has advantages in reinforcing the other strategies.

Once people get the sense that you are going to win anyway and that their true votes will not be counted, they have even less incentive to forgo the bribes and take the risks of opposition. You can also keep this one in reserve until you see that you are losing. In the Kenyan elections of December 2007, as one by one the parliamentary constituency results were declared, the opposition looked set to win the presidency. Yet by the time these constituency votes were added up to the national total by the electoral commission to determine who should be president, lo and behold, the incumbent president had narrowly won.

Cons: The international community won't like it if you push it too far. Better be a bit careful: after the Kenyan election results the European Union got upset about discrepancies. In one constituency the vote for the president had unfortunately first been announced as 50,145 before being entered as 75,261 in the final tally.

THIS ONE IS DEFINITELY FOR you. Just remember not to push it too far: not 99 percent; it should not look like a Soviet election.

So much for putting oneself in the position of the president. What struck me was how much superior, from the point of view of a self-interested political leader, some of the other options were to the tough and unreliable option of trying to be a good government. In the typical election in one of the developed countries, as defined by membership in the Organization for Economic Cooperation and Development (OECD), the incumbent government has a chance of reelection of around 45 percent. In the average election held in a society of the bottom billion, despite the fact that voters usually have many more grounds for complaint, it is a much healthier 74 percent. Political scientists have developed a scale of democratic governance called Polity IV, starting at −10, which characterizes political hell, and going right through to +10, which is political heaven. Among those countries of the bottom billion in the range −10 to zero, the president has an even healthier chance of electoral victory: an amazing 88 percent. Somehow or other, incumbents in these societies really are very good at winning elections.

I decided that it was time to investigate the winning strategies more systematically, and for this work I turned to Pedro Vicente, who already had experience from Cape Verde and São Tomé, two little islands off the coast of West Africa. I persuaded Pedro that we should be ambitious: little islands provided neat natural experiments, but we should try working on one of the major new democracies. We chose Nigeria, where elections were due during the course of 2007. Despite its evident importance as Africa's largest society, there is amazingly little quantitative field research on Nigeria. It has a reputation for being a difficult, and indeed a dangerous environment, and it is also astonishingly expensive.

All the gossip was that the Nigerian elections would be nasty. President Obasanjo had set his sights on changing the constitution so as to have a third term. The vice president, who had aspirations to the top job, set about blocking this strategy, which needed approval from the Senate. In a close and bitterly contested Senate vote the vice president succeeded in blocking the third term. This left President Obasanjo without an heir apparent of his own choosing: for obvious reasons he had not wanted there to be any alternative to himself. Worse, the vice president had entrenched himself as the likely winner, using the vice presidency to benefit from his own powers of incumbency. If there was one person President Obasanjo did not want to succeed him, it was the vice president. So, with less than twelve months before the election, he was going to have an uphill struggle to take someone from zero to victory over the vice president. As the election campaign approached he told his party it was a "do or die" affair. Everyone understood what "do or die" meant: it meant no-holds-barred. In turn, this meant "refer to the above list of options."

On one of my visits to Nigeria I had met Otive Igbuzor, an out-spoken political activist who impressed me. Although I thought that some of his views on the economy were wrong, his concerns about the lack of political accountability were cogent and passionate. He was also sufficiently open that he did not dismiss me simply because I was a foreigner. We decided to join forces. I brought a research group able to conduct a scientific field experiment; he brought a vo-cal local NGO that he headed, Action Aid, with a field network of committed people. Together we designed a field experiment to measure three of the illegitimate winning options: bribery, intimi-dation, and vote miscounting. We were also able to join forces with the team from Michigan State University that runs the Pan-African Afrobarometer survey of political attitudes. The heart of our experi-ment was to see whether voter intimidation could be countered. On a randomized basis across Nigeria, Action Aid organized powerful local campaigns against intimidation.

Manifestly, a research project aimed at trying to counter po-litical violence during a Nigerian election campaign that was an-ticipated to be particularly nasty was pushing the limits. Quite apart from the physical dangers for all the participants, Pedro had to di-vert from the safe strategy of using his time to write up his existing research for publication into this highly risky undertaking. There might easily be nothing to show for months of work, and he would need publications to get another job once the funding for his re-search post expired. Even I had to find a modicum of courage: reas-suring research foundations that they were not pouring their money down a particularly expensive drain. In the event, the elections were indeed marred by irregularities. The monitors sent by the European Union described it as "not credible," and Human Rights Watch de-scribed it as a "farce." As I write, five of the governors elected have been stripped of office by the Nigerian courts. For Nigerians the election was evidently flawed, but these very flaws made it well-suited to our research.

We found clear statistical evidence of all three strategies. The Action Aid campaign against voter intimidation had a remarkably large effect. In those randomly chosen locations in which the campaign was conducted, more people found the courage to vote. We interviewed people both before and after the election: where the campaign was conducted many more people who had initially decided not to vote changed their minds. What is more, despite this overall increase in turnout, the vote for those politicians perceived as espousing violence fell. People who had initially intended to vote for these candidates changed their minds and stayed at home.

That one campaign by one NGO could have such a big effect against such an apparently intractable problem is surely remarkable. But that was not the only surprise. We found that bribery and vote miscounting went hand in hand: they were complementary strategies. We measured them by asking people how serious they perceived bribery and ballot fraud to have been in their constituency. We found that ballot rigging favored the local incumbent party. Evidently, local incumbency is what matters for controlling the vote count. But the surprise was that voter intimidation was high when bribery and miscounting were low. It turned out that, at least in the Nigerian election, violence was predominantly a strategy of the politically weak, perhaps somewhat analogous to terrorism.

So in Nigeria politicians had clearly resorted to socially dysfunctional strategies of vote winning. Now think of the implications. With these options available, electoral competition is simply not going to deliver accountability. Nor, if politicians win by these unscrupulous means, is democracy going to confer much in the way of legitimacy. Losing opponents are not going to say, "Fair enough, you now have a mandate"; they are going to say, "You cheated" and resort to violence. In other words, democratic elections cannot possibly, *in themselves*, be a solution to the problem of violence, or to

the larger problem of decent government. In themselves they are a
recipe for driving political leadership into the gutter. It is not even a
matter of maybe. Electoral competition creates a Darwinian struggle for political survival in which the winner is the one who adopts
the most cost-effective means of attracting votes. In the absence of
restraints the most cost-effective means are simply not going to be
good governance: that option is surely way down the list.

An example from the Nigerian gubernatorial elections stared me
in the face. This had been the campaign for reelection by the incumbent
minister for the federal capital territory of Abuja, Nasir el-Rufai. Contrary to most of his colleagues, he had governed well. His ability was
recognized by ordinary Nigerians: in 2006 he had won the prestigious
Silverbird Man-of-the-Year Award. Indeed, by any standards he was
competent. He had managed to get into Harvard Business School, no
mean feat for a young Nigerian, and had duly come to the top in his
year. Also exceptionally, he had decided not to exploit the potential multiple advantages of incumbency and conducted an honest campaign. He
lost: in fact he didn't even manage to win the nomination of his own
party in the primary that preceded the gubernatorial election. Given the
potency of the dishonest options, the honest and decent have so much
stacked against them that that is all too often their fate.

So FAR I HAVE COME at this from the perspective of how to game
an election. The punch line I have been working toward is that in the
typical society of the bottom billion, electoral competition, far from
disciplining a government into good policies, drives it into worse
ones. But even if incumbent politicians resort to mischief when they
come to an election, in the meantime they might also decide to do
their best. In other words, being a good government and all the
other options may not be alternatives they may be complementary: a
scared politician may try them all. It is time to look, and for this we
need to observe not the electoral strategies but the policy choices.

Undoubtedly, during the period of electoral competition which began in the early 1990s, economic policies in the countries of the bottom billion have tended to improve. Is this causal: has democracy driven governments into better economic policies despite the mischief over how they win elections? It seems a plausible hypothesis. I had already worked on the preconditions for the reform of policies and governance with a young French economist, Lisa Chauvet. The issue of how democracy and elections affected the chances of reform was a natural extension of this earlier work, and so she was the obvious person to work with. The only problem was that she was pregnant. We raced against the arrival of little Diego to get the results I now report.

Our universe of observations was all the countries that at some stage or other had been impoverished and had had seriously dysfunctional policies and governance. From this universe, the task is to try to explain why some countries at some particular time managed to reform out of the mess, and in particular, to investigate whether democracy in general, and elections in particular, seemed to help or hinder the process. The phrase "policies and governance" is easy to write, and within reason people can agree on what they mean by it. But it is a difficult concept to measure with any precision. Further, we needed a measure of policies and governance that was available on a consistent basis for as many countries as possible, for as long a period as possible. There are only two possibilities, one put together by the World Bank, called the Country Policy and Institutional Assessment, and the other put together by a commercial company, the International Country Risk Guide. They are both based on judgments of a professional staff, a little like the process by which Standard and Poor's assigns credit ratings to country debt. We chose the World Bank rating, mainly because it started seven years earlier than the commercial rating agency and so covers a longer period.

We had already found that there were some clear preconditions that made reform easier. The larger a country's population was, the faster it reformed. I think that this is because a large population supports a market for specialized publications that can discus economic policy. India has a newspaper, *The Economic Times*, with a circulation of 1.2 million. It can afford to send its correspondents around the world. If Zambia had the equivalent with the same density of circulation, sales would be under ten thousand, and so Zambia has no *Economic Times*. It also helped to have donor technical support, the much-despised form of aid whereby skilled foreigners are sent to help governments. But now we looked at elections and democracy.

One problem with elections is that they are often not held according to a set calendar but occur due to circumstances that might themselves affect the chances of reform. Let me give you a simple example of how an unsuspecting researcher could get into trouble. Suppose that what is really going on is that periodically the people pressing for change in their society manage to break through politically. They believe in economic reform and they also believe in democracy. So they hold an election and they also reform the economy. If the researcher is not careful, this is going to look as if elections cause reform. So how can the researcher be careful? The answer is to find something that is a reasonable predictor of when the next election will be held but that does not itself influence current prospects of reform. The best we could think of for this was to predict the timing of the election on the basis of the time lapse between the two preceding elections. The idea is that in many societies there is a fairly fixed frequency of elections. Indeed in some it is set in concrete, as in America. Repeating the analysis just using these countries where the government cannot choose the election date is a simple way of checking the reliability of the results.

We then asked how the amount of time left until the next election affected whether policy and governance improve or worsen. We found a clear and unambiguous relationship. For the first few years after an election the chances of policy improvement got better, year by year. And then, as the next election approached, the chances of reform started to get worse again, year by year. Two years before the election the chances of reform slipped, and the year before the election reform was highly unlikely. What the results were telling us was that the chances of reform were at their peak when the society was as far away from an election—in either direction—as possible. Why might this be? Perhaps, in the first year or so after an election, the government was too new to be able to implement reform, and as the election approached it was too preoccupied by the need to win the election to bother with reform. After all, the payoff to most reforms takes several years, and any payoff that does not arrive until after the election has little political benefit.

This was not really encouraging: it suggested that elections were to an extent a distraction rather than a stimulus. I recalled my friend Ngozi Nkonjo-Iweala telling me when she became Nigeria's finance minister that although the government was at the start of a four-year term, she had been given only three years for reform. "The last year will be politics," the president had explained to her, and, as I have just described, so it had proved. However, it might nevertheless be that all the election effect showed was a variant of the political business cycle. The political business cycle was the game that rich-country politicians used to play with their own electorates, pumping money into the economy just before an election: whoever won would then have to clean up the mess in the next couple of years. Damaging as the political business cycle was, it did not mean that democracy was worse than autocracy. It just showed that it wasn't perfect. So the election results in themselves did not say anything about whether, if your society needed reform, democracy was better or worse than autocracy.

To investigate this deeper question, Lisa and I then introduced measures of the polity. How democratic was it? Was government power limited by checks and balances? In particular, were elections well conducted? Fortunately, all these characteristics are now classified and coded by political scientists. For example, we used a standard measure known as Polity IV, which you have already come across. This arrays the degree of democracy on a scale from 0 to 10, alongside a scale for autocracy, which is ranged from −10 to 0. So the people's paradise of North Korea scores −10, the squeaky clean democracies of Norway and Switzerland score +10, and a messy electoral competition typical of the bottom billion would score at best around +2 or +3. Prior to the wave of democratization the societies of the bottom billion had on average been around −6: in other words, they were mostly autocracies. Currently, the average score is around zero. When we added these characteristics they mattered: the electoral cycle overlaid, and potentially confused, these deeper effects. Elections can potentially spur a government to adopt reforms, but they can also drag it further down the road to bad governance. Which effect predominates depends partly upon structural features of the society, and partly upon the design of the polity. Elections tend to work better in societies that have larger populations and fewer ethnic divisions. They also tend to work better in polities with checks and balances on the power of government, and in particular where the elections are properly conducted. On the evidence, elections without properly enforced rules of conduct in small, ethnically divided societies typically retard reform rather than accelerate it.

So the implication is that to date the process of democratization in the bottom billion has remained within the range over which better is worse: the increased democracy has quite probably retarded the reform of economic policies and governance. It has gone far enough to lose whatever might be the advantages of autocracy, while not yet having gone far enough to gain the benefits of democracy, and the typical society of the bottom billion remains well short of the point at which democratization would lead to improvement. It has proved much easier to introduce elections than checks and balances. Presidents quite enjoy being anointed by the holy oil of an electoral victory, whereas they find the prospect of effective checks and balances truly alarming. But above all, they have woken up to implications of the lack of checks and balances for their ability to survive elections.

Taken together, the results on elections and democratization are consistent: if democracy means little more than elections, it is damaging to the reform process. I do not like these results. It would be a much happier story if at every step along the way to fully fledged democracy the consequences got better and better. But unfortunately this does not seem to be the world as it is.

The results on the dysfunctional consequences of partial democracy for reform are also consistent with the evidence on how elections are actually won in the societies of the bottom billion. The six nefarious options for winning an election not only dominate the option of trying to be a good government, collectively they constitute an alternative. So why don't more governments hedge their bets and do both: win by foul means, but improve their chances that extra bit by also trying to be a good government? I think it is because *the other options depend upon bad governance*. If you want to use them you have to sacrifice the strategy of being a good government even if you recognize that otherwise it might be worth doing.

One reason for the conflict between decent governance and the other options is money. When President Obasanjo realized that he would not be able to stand for a third term, he knew that he was in for a very tough contest. How do you win a Nigerian election for an unknown candidate in only a few months, facing an entrenched opponent? The answer is you probably need a lot of money. Yet over the previous three years President Obasanjo had started to put in place the rudiments of accountable government finances. He had entrusted the ministry of finance to Ngozi Nkonjo-Iweala, and public procurement to Oby Ezekwesili. These two tough, able, Christian women had shut down the sources for the sort of slush money that a political campaign was likely to need. Within a month of the Senate decision to deny President Obasanjo a third term, he had shifted both of them well away from control of government money. The one high-profile fighter against corruption who was not shifted was Nuhu Ribadu, head of the Economic and Financial Crimes Commission. Bravely, in late 2007, he launched a prosecution against James Ibori, the key financial backer of President Obasanjo's chosen successor. Ribadu lasted only a further three months before being ousted.

The more effective strategies are also incompatible with the rule of law. When President Mugabe discovered that he had lost the referendum on removing the term limit, and so realized that he would lose the next election, he set about the process of dismantling the rule of law, starting by forcing the chief justice into early retirement and appointing a placeman. As the rule of law was gradually lifted, new options opened up for snatching revenues at the expense of the economy and President Mugabe duly took them: property rights were ignored, and finally he resorted to hyperinflation. In other words, the government needs to remove checks and balances in order to use the other electoral options, and with checks and balances removed, other policies are very likely to deteriorate.

This is, unfortunately, consistent with some new work by Masayuki Kudamatsu that carefully tries to investigate whether the introduction of elections in Africa has led to a reduction in infant mortality. Reducing infant mortality is surely about the most basic concern of ordinary citizens, and across the bottom billion, infant mortality has been avoidably high. An election should surely empower citizens to force governments to reduce the risk that their young children will die. He concludes that only following those rare elections in which the incumbent president was defeated did infant mortality fall. In the more normal situation of incumbent power, elections achieved nothing.

So both the evidence on how elections are actually won, and the actual policy performance of democratic governments in the bottom billion, point to the same conclusion: in the conditions of the bottom billion, electoral competition is not producing accountable government. I started by noting that there had been a considerable improvement in economic policies and governance across the bottom billion coincident with the spread of electoral competition. So, if elections have not caused the improvement, what has?

I think there are two likely explanations. The simpler, and therefore probably the better, is that societies have learned from past mistakes. Experience is a hard school, but all the societies of the bottom billion have been through it. The high-income world has obviously learned from its mistakes: the inflation of the 1970s is a thing of the past because electorates in the high-income societies will no longer put up with it and governments have learned how to tame it. Quite probably the same process has been going on in Africa. With the exception of Zimbabwe, inflation rates are now far lower than they used to be. Whether or not electorates in the bottom billion have much influence on their governments, elites may well have woken up to the fact that inflation and some other dysfunctional economic policies are not worthwhile.

The other possible explanation is that donor conditionality has imposed discipline on governments, forcing them into reform even if they did not want to do it. I do not entirely discount this explanation. It is very difficult to sort out the motivations behind actions. To an extent donor conditionality might have forced reform. But the statistical evidence if anything suggests that it has delayed reform rather than accelerated it. Governments do not like being made to do something against their will and they are remarkably ingenious at finding ways of not doing it. Donors are also amazingly bad at enforcing their agreements with governments. So my own judgment is that donor conditionality on economic policies is not the explanation for policy improvement. I would put my money on learning from failure.

I REALIZED THAT IF THIS critique of electoral competition was right it had huge implications. The whole modern approach to-

ward failing states had been based on the premise that they would
be rescued by democratic elections. The approach had seemed to be
vindicated by the enthusiastic take-up of elections even in the most
unpromising circumstances. Afghanistan, among the most backward societies on earth, was able to run an election within months
of the expulsion of the Taliban. Iraq, about the most violence-torn
place on earth, was able to conduct an election with quite a high
turnout. The Democratic Republic of the Congo, a society with the
staggering misfortune of Belgian colonialism, followed by Mobutu,
followed by civil war, was still able to hold a competitive election.
The dread shown by the Soviet authorities to any form of competitive election has, I think, confused us into thinking that achieving a
competitive election is in itself the key triumph. The reality is that
rigging elections is not daunting: only the truly paranoid dictators
avoid them.

Why is it so easy to hold elections even in unpromising circumstances? Surely it is because both political parties and voters
face strong incentives to participate in them. For political parties the incentive is that the election is the route to power. For
the governing party it is a fair bet that the election will consoli-
date power and gain legitimacy in the eyes of the donors. For
opposition parties there is at least a chance of power, and with
the governing party mobilizing its supporters, even if victory is
unlikely it is important to have a countermobilization of support, otherwise it will drain away. Why do voters bother to vote?
Economists have tied themselves in knots here, missing the obvious. We are so wedded to the notion that people's actions must be
in their material self-interest in order to be rational that our approach is largely confined to what is known as instrumentalist—
or, more colloquially, "What's in it for me?" A young Northern
Irish economist at Oxford, Colin Jennings, helped me to think on
more realistic lines. Obviously influenced by his Northern Irish
experience, he emphasized the satisfaction that people get from

using their vote as a way of expressing their identity: voting is satisfying in the same way that wearing a football scarf is satisfying. And so voter turnout is likely to be particularly high where identity politics rules. Paradoxically, the less politics is about policies, the stuff of instrumental voting theories, the stronger is people's incentive to vote. In America voting may be instrumental. Indeed, perhaps that helps to explain the low turnout; but in the divided societies of the bottom billion, voting is likely to be primarily expressive.

IT IS TIME TO SUM up where we have got to, and it is not attractive. Democracy, at least in the form it has usually taken to date in the societies of the bottom billion, does not seem to enhance the prospects of internal peace. On the contrary, it seems to increase proneness to political violence. Probably related to this failure to secure social peace, democracy has not yet produced accountable and therefore legitimate government.

Incumbent politicians have won elections by methods that require them to misgovern. This is supported by the evidence that democracy seems to retard reform.

In promoting elections the rich, liberal democracies have basically missed the point. We want to make the bottom billion look like us, but we forget how we got to where we now are. We did not do it in a single leap: dictatorship to liberal democracy. We have been unrealistic in expecting that these societies could in one step make a transition that historically has been made in several distinct steps.

Perhaps, in encouraging elections, we have landed these societies in an unviable halfway house that has neither the capacity of autocracies to act decisively nor the accountability of a genuine democracy. Soon I am going to argue that it is not as hopeless as it might appear. But we have not yet done with the upsetting material:

the story is going to get worse before it gets better. I will close with
the comment, made the day after the Kenyan election by Michael
Ranneberger, the American ambassador. "It's a sad day for Kenya,"
he lamented, but then came the acid: "My biggest worry now is violence, which, let's be honest, will be along tribal lines."*

---

* "Tribal Rivalry Boils Over in Kenyan Election," *New York Times*, December 30, 2007.

Chapter 2

ETHNIC POLITICS

IN THOSE KENYAN ELECTIONS THE opposition candidate, Raila Odinga, was a Luo, one of Kenya's forty-eight ethnic groups. He secured 98 percent of the Luo vote. This was identity voting with a vengeance. Does it matter?

Everyone has some subnational identity and usually several of them: in addition to being British I am English, more particularly *Northern* English, and getting right to the bone, I am a Yorkshireman; I have taught my son Daniel to sing our anthem, "On Ilkley Moor bar t'at." The British general election of 2001 pitched a Yorkshireman against a Scot for prime minister. Yet like most people from Yorkshire I supported the Scot. A society can function perfectly well if its citizens hold multiple identities, but problems arise when those subnational identities arouse loyalties that override loyalty to the nation as a whole. As the Luo vote suggests, in the societies of the bottom billion, ethnic identity usually trumps national identity.

The societies of the bottom billion are for the most part far more ethnically diverse than those of the high-income countries. Often this diversity verges on being a taboo subject: it is just too upsetting. I think that it poses genuinely tough, but not insuperable, problems. They will not be overcome unless they are faced.

Ethnic diversity compounds the problems that the societies of the bottom billion would in any case face in making electoral competition work. Yet more fundamentally, diversity impedes the basic role of the state, the provision of public goods. It is tempting to conclude that what an ethnically diverse society of the bottom billion needs is a strongman. Wrong: bad as democracy is in ethnically diverse societies, dictators are even worse. But there is a vital role for political leadership: leaders must build the nation before they can build the state.

WHAT WAS THE ORIGIN OF these strong ethnic loyalties? In the absence of states, ethnicity was the obvious basis for collective action, and in a rural society bumping along at a subsistence level of income, one form of collective action was supremely important: insurance. Life at subsistence is risky: if you fall sick when you should be plowing, planting, or harvesting, your income will collapse. If vermin eat your food stores, you face starvation. You need catastrophe insurance. The problem with insurance is what economists coyly term moral hazard: if I'm insured, what the heck! If you could insure yourself against a decline in income, why get up in the morning? And so such insurance does not exist unless the moral hazard problem can be solved. The solution to moral hazard is not indignantly to protest that the insurer should not doubt your good faith, it is to make your behavior observable. Only if the insurer can see that you are trying your best does the insurance become feasible. For a private insurance company such observation would be prohibitively expensive, but for a community it is feasible. Nosiness, gossip, friendly intimacy, all the ingredients that are natural to a community also happen to be just what is needed for insurance.

Observability is necessary but not sufficient. The right to rely upon other people in the community when faced by a personal catastrophe depends upon a reciprocal obligation to provide such assistance to others: but who is in and who is out? If anyone can join or leave the insurance group at any time, then it will be in perpetual deficit: people will declare themselves to be members of the community when they fall on hard times and declare themselves fancy-free when things are going well. This is known in economics as the problem of adverse selection: unless insurance companies take care, instead of getting a random selection of clients from the population, they get people who know that they are bad risks. That is why insurance companies use some device for restoring a random selection, such as offering much better terms for all the employees of a firm than they offer to individuals who turn up at the door. This is where ethnicity comes in: you do not choose your ethnic group. If you are not a member of the ethnic group, you cannot choose to become a member when times are hard. If you are a member, you cannot choose to exit the group when things go well. That is the economic basis for strong ethnic loyalties: it enables income insurance to work in the high-risk, low-income conditions under which it is supremely valuable. Over time, loyalty to the group becomes reinforced by all the normal power of morality: it is morally good to meet your obligations.

Insurance sustained by loyalty helps everyone within the group and is not at the expense of other groups. However, even in the traditional economy loyalty to the group is sometimes at the expense of other groups, most obviously in respect of violence against enemy groups. But ethnic loyalties have far more scope for being at the expense of other groups when they are transferred to the context of the modern economy. The public purse becomes the common pool resource that the collective action of one group can capture at the expense of other groups. It is at this stage that moral obligations to the ethnic group collide with moral obligations to society as a whole.

My friend John Githongo, anticorruption commissioner in the Kenyan government, blew the whistle on the corruption at the heart of the government, becoming internationally famous and an exile in the process. Even I could see that what John did took courage.

But in talking to him I got a surprise: it had taken more than courage. John is a Kikuyu, the tribe that dominates the government. Unsurprisingly, when he blew the whistle against the government, his Kikuyu friends had accused him of betrayal. But the surprise was that John had felt this himself as an inner struggle of conflicting loyalties. Like many of the finest African reformers, John is a committed Christian: religious faith gives a moral framework that helps people to put their ethnic obligations in perspective. Another committed Christian reformer, Oby Ezekwesili, who bravely slammed the door shut on Nigeria's public procurement scams, described the prevailing morality: "They see themselves as good if they benefit a few thousand kin at the expense of the nation." It is that prevailing morality that gives ethnic loyalties a whole new possibility for being valuable to the group yet damaging for the society.

Far from the transition to the modern economy weakening ethnic bonds, there are strong forces intensifying them. Occasionally an event gets under the skin of a society and reveals much more than its direct importance. Here is the story of a Kenyan funeral. Like Raila Odinga, Mr. Otieno was a Luo. However, he had left his native region in his youth and moved to Nairobi, where he had become a successful businessman and married a Kikuyu woman. So far we have a standard story of the melting pot. On his death in 1986, his widow, in accordance with Mr. Otieno's last will and testament, arranged for his burial in Nairobi. At this point Mr. Otieno's Luo relatives objected: they wanted him buried back home. Indeed, they wanted him home so badly that they took the matter to court. Faced with a choice between adhering to the wishes of the dead man and his wife, or the wishes of the relatives, the court was in no doubt: he was duly buried back in his Luo village.

What on earth was going on? Think back to the key policing roles of ethnicity: entry and exit from the obligations of the group. There is little difficulty in policing entry: "Very sorry but we won't help you because you're not one of us." But policing exit is rather harder. The people who will want to exit from their obligations are the successful: and how do you stop them? This is where the place of burial comes in. The spirits of ancestors loom large in the belief systems of most ancient societies, and spirits are usually localized. Mr. Otieno might have managed to exit his obligations during life, but he might now be getting his comeuppance in death. Consciously, or subliminally, an enforcement mechanism for ethnic loyalty was at work; and unlike Mr. Otieno it was alive and well.

SO WHAT HAPPENS IF THERE are many ethnic groups, each with powerful loyalties? How does it affect the politics?

Electoral competition is an activity with powerful economies of scale: if I can get 51 percent of the votes I win. Indeed, in the absence of restraints on the use of power, I win everything. To reap these economies of scale, power seekers group together into political parties that develop brands and try to build voter loyalty. In ethnically homogenous societies with winner-take-all voting systems, this process tends to be driven to the extreme in which everyone amalgamates into only two parties. Although the leaders of these parties are chosen only by their respective supporters, once chosen, both leaders chase the median voter to get elected. This produces a politics of moderation that broadly describes how modern democracies function. One hallmark is that the activists within each political party are usually dissatisfied with the moderation of their leaders. We see the process played out most clearly in America, but with minor variations it is the general pattern.

When I first tried to work out how this process was affected by ethnic diversity I came away quite heartened. Of course, if voters had strong ethnic identities then politicians would organize their parties on ethnic lines: this would simply be the cheapest way of attracting voter loyalty. The election itself would sound very different from an election in an ethnically homogenous society: leaders would simply be mobilizing their own ethnic base rather than reaching out to the median voter. But after the election these ethnic parties would need to form coalitions. Any ethnic group that got too demanding would not be able to bid itself into the winning coalition. Ethnic politics might produce a merry-go-round of changes in governments, but each group would get close to its fair share of power.

Since publishing those ideas in 2001 I have started to have doubts. First, ethnic politics seems likely to contaminate the content of the election campaign. Policy choices get crowded out by identity. Let's go back to those winning electoral strategies: remember the one that involved playing the ethnic card. Playing on ethnic fears and hatreds is truly the politics of the gutter: unfortunately, it works. The holy grail of modern economic field research is the randomized experiment, something that medics have been doing for years, but it is usually more difficult to arrange with economic interventions. When it comes to the content of an election campaign, you would imagine that the scope for conducting a genuinely randomized experiment is decidedly limited. Not one bit of it: Leonard Wantchekon, a remarkable economist from Benin now working in America, did just that. He managed to persuade the politicians of Benin randomly to adopt different campaign messages in different localities. This alone tells you most of what you need to know about the election campaign in Benin, but Wantchekon's story is yet more depressing. Not only were politicians willing randomly to adopt either a campaign message that they would provide good national governance or a message that they would provide ethnic favoritism, but once the results were subject to statistical analysis, it became clear that favoritism was more effective at pulling in the votes.

Not only does identity trump policies, but to the extent that policies do enter, instead of a race to capture the vote of Ms. Moderate the All-Powerful Median Voter, there is a race to the extremes. Colin Jennings introduced me to this tendency via the expressive voting idea. His work analyzes how electoral competition is likely to work out in ethnically divided societies. Voting for the extremist parties offers the strongest identity fix. It also selects the most ardently sectarian leaders, so that when it comes to the stage of reaching compromise in a grand coalition, the starting point for the negotiations is as far toward the position of your own ethnic group as possible.

One graphic instance of this unattractive process is evident in Northern Ireland, where electoral competition was meant to force moderation, with parties heading for the center ground in order to build a coalition. Instead, precisely the opposite happened. There are four major political parties in Northern Ireland, two Protestant and two Catholic. On each side one of these parties is moderate and one is extreme. Prior to power sharing, the largest parties on each side of the Protestant-Catholic divide were the moderate parties: indeed, they were the parties that brokered the power sharing. But once power sharing was introduced, voters polarized; now the dominant parties on each side are the extremists. The ruling coalition is a coalition of the extremes headed by grinning bigots who cannot believe their luck. This seems to be a likely consequence of identity politics more generally. Indeed, it happened in the Kenyan elections of December 2007. The forty-eight ethnic groups coalesced into pro-Kikuyu and anti-Kikuyu coalitions.

I also came to see that electoral competition is not the only aspect of democracy that matters. Electoral competition needs to be complemented by checks and balances. In turn, checks and balances are public goods: that is, they have to be supplied by cooperation. Ethnic politics makes such cooperation to build checks and balances much harder. I came across this graphically in the aftermath of the Nigerian elections of 2007. The new speaker of the House of Representatives, Patricia Etteh, was soon caught misappropriating the funds meant for her office. She had, for example, acquired twelve Mercedes. I do not want to get Scandinavian about the odd dozen Mercedes: I am quite prepared to believe that any self-respecting speaker needs them. But many Nigerians seemed not to think so: they regarded it as outrageous and she was pilloried in the media. So far there is nothing remarkable in this story: a relatively minor infringement that met its fate. It is the reaction to the criticism that is of significance. As soon as she was criticized in the media, the other politicians from her own ethnic group, the Yoruba, leaped to her defense. Quite explicitly, their message was "Hands off: she's our only representative at the trough." If a corruption charge can be deflected by playing the ethnic card, then standards of public conduct are bound to be low.

So ethnic electoral politics may not be as benign as I had previously thought. This would certainly gel with the evidence on ethnic diversity and public goods, much of which is derived from contexts in which political choices are the result of vigorous electoral competition, such as North American cities.

MANY STUDIES HAVE FOUND THAT public services are systematically worse as a result of ethnic diversity among citizens. The association is causal: it is not just that ethnically diverse societies happen also to have poor public services. Controlling for other characteristics, greater diversity implies worse public services. Not only that, but expenditure on channels suited to ethnic patronage, such as the public payroll, is higher. Why does diversity make public goods provision harder? For that we need to turn to micro-level evidence on how collective decisions are taken.

One result clearly established by research is that trust is weaker across ethnic groups than within them. One rather clever way of demonstrating this was the work of Abigail Barr, a researcher in my group, who investigated variations in the level of trust among the communities of rural Zimbabwe. Trust is difficult to measure, but she followed a recent line of research and used experimental games played by volunteers who could win small amounts of money depending upon the strategies they adopted. Zimbabwe was particularly well suited to the investigation because alongside ethnically homogenous villages were others that had been created as settlement schemes at various times, in which there were varying degrees of ethnic mix. She was able to show that controlling for other characteristics, the more ethnically mixed villages revealed a lower level of trust through the strategies people chose to play. Another result is that people are more willing to pay taxes to benefit people with whom they have an affinity than if they know that much of the expenditure will benefit people who are very different. When these results were reported in Europe, they produced a frisson of concern that immigration and the resulting change to multiethnic societies would erode the welfare states that characterize the continent.

There is also evidence that the public good of scrutiny of government breaks down. I have already recounted the anecdote about the Mercedes bought for the speaker of the Nigerian House, but there is more systematic evidence. A particularly convincing study compares the functioning of school boards in various parts of rural Kenya. The school boards, composed of parents, can raise funds and manage the school: they thus have an important role to play in determining school quality. A clever study by Edward Miguel and Mary Kay Gugerty found that where the board was diverse, management was worse: specifically, the members of the board were not prepared to criticize people from their own ethnic group who were failing to make contributions.

Fortunately, there is a silver lining to ethnic diversity. The adverse effect of diversity on public provision is offset by an advantage that it confers in private economic activity. Why does diversity enhance private sector productivity? There is now pretty good experimental evidence that reveals what is going on, somewhat along the lines of Abigail's work in Zimbabwe. Basically, diversity raises the productivity of a team because it increases the range of skills, knowledge, and perspectives, and these help problem solving. Although diverse teams do not get along as well, they are better at achieving results. There is also some evidence that suggests that this scales up so as to affect the overall performance of an economy. I am not particularly proud of my own effort in this direction: I pushed to the limits of data availability, and so the results are probably not that solid. But, for what it is worth, I constructed estimates, country by country, of the public and private capital stock and then investigated whether the productivity of these two types of capital was affected by the degree of ethnic diversity of the society. Each of these steps is precarious, but what came out was that ethnic diversity reduced the productivity of public capital, while increasing the productivity of private capital. While this may be spurious, it is at least consistent both with the micro-level evidence and with other macro-level results.

One implication is that diverse societies should play to their advantage and place as many activities as possible in the private sector. This is clearly consistent with the contrast between America and Europe: the as yet more homogenous societies of Europe have a larger public sector. The societies of the bottom billion, with their high diversity, were particularly ill suited to the socialism that until recently has been overwhelmingly their predominant ideology. Their adoption of socialism was understandable: most of the first generation of political leaders had been educated in France and Britain in the 1950s. Not only had socialism been at its apogee, but to their great credit Europe's socialists were the first politicians to support decolonization struggles. And beyond European socialism, imitating the Soviet model carried the sweetener of a ready access to armaments to address their security problems. One aspect of the so-called Structural Adjustment Programs of the 1980s was that African governments were encouraged or coerced into shifting activities from the public sector to the private sector. Though heavily criticized both for being coercive and for being ideologically driven, the direction of the shift was appropriate given the diverse composition of Africa's societies.

Since diversity has beneficial effects as well as adverse ones, it sounds as though with appropriate choices it might be a case of six of one and a half dozen of the other. The net effect might be negligible. However, you have already seen that an effect can be very different in high- and low-income societies. Recall that democracy reduces political violence in high-income societies but increases it in low-income societies. Could the effect of ethnic diversity be similar?

Unfortunately for the societies of the bottom billion, it is. The beneficial effects of diversity only set in at higher levels of income: diversity is good news for America, and while Europe's rising diversity may well weaken its welfare state, it will be compensated by a more vibrant private economy. But it is bad news for Kenya and the other societies of the bottom billion. At low levels of income, diversity is a substantial net economic disadvantage, and it shows up in slower growth: a highly diverse low-income society on average grows a full 2 percentage points less rapidly that a completely homogenous one. Why might diversity help high-income societies but hinder low-income societies? Perhaps it is because the key advantages of diversity come from skills and knowledge. In an economy with high levels of skills and knowledge, the larger the pool of diverse skills and knowledge, the better. But in economies where skills and knowledge are more rudimentary, there is less scope for diversity and less use for it.

FUNDAMENTALLY, THE RESULTS SO FAR suggest that ethnic diversity makes social cooperation more difficult, and that at low income levels this effect is sufficiently strong to be a substantial impediment to prosperity. It is tempting to conclude from this that diverse societies cannot afford to rely upon cooperation to achieve the collective effort that is necessary for success in any economy. The alternative to achieving collective effort through cooperation is to achieve it through coercion. Someone is needed to direct the coercion: step forward the benign dictator. The recent evidence of China illustrates a more widespread phenomenon, that a society can make rapid economic progress if collective effort is guided by a sufficiently sensible and relatively benign autocratic leadership. Is this the answer for ethnically diverse low-income societies?

The case for autocracy appears to be strengthened once we get back to the fundamental issue of security. You have already seen that in the bottom billion, democracy increases political violence in all its main forms. While democracy makes such societies more dangerous, repression seems to work. We are back with the ugly fact that Saddam Hussein kept the peace in Iraq more effectively than Jalal Talabani. So better public goods—dictators make the trains run on time—and better security: the case for dictatorship looks disturbingly strong.

While I do not want to discount the benefits from a sensible and benign autocrat, I think that for ethnically diverse societies, this solution to the problem of collective action is very dangerous. Ethnic diversity generates bad autocracies as well as bad democracies. In an ethnically diverse society dictators usually play the ethnic card, building their power base on their own ethnic group. As a result, their patronage base is almost inevitably narrow, not extending beyond their ethnic group. The narrower the power base, the stronger is the incentive to retain power by raping the national economy and transferring the proceeds to their own ethnic group rather than by building the national economy and benefiting everyone. So, on this analysis, ethnically diverse societies would be supremely ill suited to dictatorship.

Again, it is best to look at the evidence, but this particular question is far from straightforward. I started with a rough-and-ready attempt. What I found was that, judged by economic performance, ethnically diverse societies needed democracy more than those that were homogenous. If this result was correct, then far from needing a dictator, ethnically diverse societies were peculiarly ill suited to them. While the result was sufficiently new to get published in a respectable academic journal, it was manifestly only a first step at an important question and quite possibly a misleading one. Recently Eliana La Ferrara and her distinguished coauthor Alberto Alesina, head of the economics department at Harvard, have revisited the issue and published a considerably more thorough analysis. I read their study with the mixed emotions of delight that the topic had engaged such a heavyweight team, admiration that they had pursued possibilities that I had missed, and, of course, trepidation that my own work might be revealed as a house of cards. In the academic world you are never more than one demolishing article away from humiliation.

One important result that I had missed was that diversity was less damaging at higher levels of income. But potentially this spelled the death knell for my own result on democracy. Since democracy is more common at higher income, my result, which did not control for the level of income, may simply have been due to this correlation. They built up their analysis step by step, first replicating my result, then revealing their own, and finally combining the two possibilities. Happily both for ethnically diverse democracies and for my own peace of mind, they found that both effects survived: ethnic diversity is less problematic at higher levels of income, and is differentially well suited to democracy. Even their analysis is only preliminary. As they acknowledge, there are various ways in which these results could be spurious. Nevertheless, the results caution against the leap from the problems of ethnic democratic politics to the inference that what is needed is a dictator.

Following their work I have tried to take the analysis a little further. Both my previous study and that of Alesina and La Ferrara took economic growth as the measure of performance. For some purposes this is not a bad measure. If there are indeed two offsetting effects of diversity, the key issue is whether the net effect is positive or negative. As a composite measure of performance, growth is as good as any. However, if, as seems likely, for the societies of the bottom billion the net effect is indeed negative, then we need to drill down. The adverse effects on public goods that are doing the damage must run through political or social choices. I therefore decided to switch from looking at the effects on growth to a more direct measure of these choices.

The issue was evidently on the boundary between economics and politics, and I was fortunate to be able to team up with Robert Bates, like Alesina a professor at Harvard, and the undisputed doyen of political scientists working on Africa. For some years we have been part of a large team under the auspices of an African-directed research network. The team had been investigating the fraught topic of why Africa's economies had largely stagnated during the forty years after 1960. Poor choices were by no means the only explanation for stagnation. For example, the fact that so many African countries were landlocked was a fundamental impediment to prosperity. However, choices were clearly contributing factors, and the team decided to focus on those that were manifestly dysfunctional. We met up at Stanford one summer and went through the narratives of economic history, country by country. What emerged was a consensus on a few syndromes. One, for example, was the mismanagement of booms, gearing them up by borrowing and then squandering the proceeds. We found that where countries stayed clear of these syndromes they always avoided economic collapse, even if they did not grow rapidly.

Bates and I decided to use these killer syndromes as our measure of performance. Did ethnic diversity make a country more prone to these highly dysfunctional choices? We found that the ruinous combination was high diversity together with severe political repression. This was the cocktail that had produced Africa's dysfunctional social choices. Indeed, it was only through this lethal interaction that diversity and dictatorship made a society more prone to the syndromes. It was not dictatorship in itself, or diversity in itself, but only their combination. This result, which is entirely based on variation among African societies, is clearly consistent with the globally based results: ethnically diverse low-income societies are particularly ill suited to autocracy.

Finally I turn to what is to my mind the most insightful study: an as yet unpublished paper by Tim Besley and his student Masayuki Kudamatsu, provocatively entitled "Making Autocracy Work." They show that performance in autocracies is far more dispersed than that in democracies: autocracies can be extremely successful but also utterly ruinous. Their question is what drives the difference: why were none of the successful autocracies in Africa? They build their answer around the notion of a selectariat. A selectariat is what a dictatorship has instead of an electorate: it is the limited group of people on whom power rests. These are the people who could therefore potentially oust the dictator if he performs badly. Besley and Kudamatsu discovered that the difference between successful and ruinous dictatorships is whether the selectariat is willing to use this power. Where selectariats routinely ditch incompetent dictators, the autocracy performs well.

This result is important, but it raises a further question: what determines whether the selectariat is willing to ditch a failing dictator? They come up with a simple answer. The selectariat will only dump the dictator if it is confident of retaining power, replacing him with one of their own. Here, I think, lies the explanation for why autocracy goes wrong in societies stratified by strong ethnic identities: in such societies political change is risky. The current selectariat will be drawn from the ethnic group of the dictator, but if it ousts him, it may trigger a chain of events in which power passes to a rival ethnic group and thus to a new selectariat. When I discuss coups you will find evidence consistent with this: in Africa ethnic polarization strongly increases the risk of a coup. An ethnic selectariat would be right to be fearful of disturbing the status quo. Consistent with this argument, Besley and Kudamatsu find that ethnic diversity reduces the chances that an autocracy will work. But they also find evidence that ethnic diversity is far from being the whole story: its effects can be overridden. A strong ideology such as Marxism makes autocracy more likely to succeed even in the context of ethnic diversity. If the selectariat consists of the Communist Party, whoever heads the dictatorship, the party is going to remain in power. The societies of the bottom billion do not need another dose of Marxism. But they do need something that gives a sense of common identity.

So, while neither economic theory nor statistical analysis has yet been able decisively to nail the issue, as far as we can tell it looks as though the tough autocrat who rules by fear is precisely what the diverse societies of the bottom billion most need to avoid. Although they are able to keep the lid on political violence other than their own, measured on a wider array of criteria, they are a disaster. Diversity may make democratic politics deteriorate, but it is likely to make dictatorship lethal.

So HOW CAN ETHNIC DIVERSITY be overcome? A sense of national identity does not grow out of the soil: it is constructed by political leadership. A few political leaders of low-income societies have succeeded in countering the problems posed by ethnic diversity by superimposing a constructed national identity. Two outstanding instances were Sukarno, who was president of Indonesia from 1945 until 1967, and Julius Nyerere, who was president of Tanzania from 1964 until 1985. More recently Nelson Mandela set South Africa on the same path. Both Sukarno and Nyerere got their economic policies seriously wrong, falling victim to the fashionable nostrums of their times, but on the key issue of building the nation they were political giants. Sukarno had the more difficult task, a vast territory of more than six thousand inhabited islands.

This has indeed always been how national identity comes about: it is a political construction. But here I want to stick with the rare instances of the construction of a sense of nation in the new post-colonial countries. What can leaders do?

Both Sukarno and Nyerere focused on language: indeed, language is so fundamental to ethnic identification that it is the main way in which social scientists have measured it. Sukarno created a national language, Bahasa Indonesia, so simple that I have heard Australian schoolchildren chatting away confidently in it. Nyerere made Kiswahili universal across Tanzania. From now on I am going to focus on Nyerere's strategy, for reasons that will soon become clear.

Language was not the only strategy for surmounting tribal identity that he adopted. He took charge of the primary school curriculum, inserting a heavy dose of pan-Tanzanian history into it. Children were taught in school to see themselves as Tanzanians. While language and education policies tried to reshape cultural identity, Nyerere also transformed the processes whereby political decisions were taken. He eschewed multiparty electoral competition, sensing that it would be divisive. Instead, at the local level the colonial system of enhancing the power of the tribal chief was completely uprooted. The national political party created village committees. At the national level resources were allocated between localities, and hence between ethnic groups, on principles of equity. Nyerere also constructed physical symbols of national unity, most notably building a new national capital, Dodoma, in the center of the country, an act much derided by the donors. Partly due to lack of funding, Dodoma has not succeeded, but it clearly demonstrated his larger purpose of moving beyond the inherited localized identities. Above all, Nyerere developed and hammered home the rhetoric of national unity: people were Tanzanians, and that was something to be proud of. Ethnic identities were not forcibly suppressed; they were simply downplayed. Even when Tanzania introduced multiparty politics it was circumscribed: no party was allowed to campaign on an ethnic platform. By chance, the current leader of the Tanzanian opposition is an old friend of mine: a fine economist, he is about as far removed from the politics of the gutter as it is possible to get.

Did Nyerere's strategy work? That is one of those questions that it is intrinsically difficult to approach scientifically. One guide is the Afrobarometer survey, which asked the same attitudinal questions in many African countries. One of the questions got pretty close to the heart of identity: it asked, "Which specific group do you belong to first and foremost?" and the potential responses were left open-ended. Elsewhere in ethnically diverse African societies, nearly half the responses were couched in terms of ethnicity: first and foremost people defined themselves in ethnic terms. In Tanzania only 3 percent responded with an ethnic or linguistic identifier. Having to identify themselves more specifically than simply "Tanzanian," three-quarters gave their occupation. I think I would do the same: proud as I am of my origins, I identify myself more strongly as an economist than as a Yorkshireman.

But these responses to survey questions may reveal no more than what is deemed acceptable in polite discourse: people may reply to the interviewer by saying whatever makes them look good. Economists are generally rather suspicious of reaching conclusions about behavior just on what people say about themselves; we prefer to infer true opinions from what people do. So the real issue is whether differences in the sense of identity drive differences in behavior. This question is more difficult. Difficult, but not, as it happens, impossible: Edward Miguel of Berkeley recently did it. This is how.

Nyerere's attempt at nation building in Tanzania stands in stark contrast to political leadership in neighboring Kenya. Kenya's first president, Jomo Kenyatta, was in many respects also a great man: his economic policies were far better than Nyerere's. When Tanzanian socialists accused Kenyatta of running a "man-eats-man" society, Kenyans aptly responded that Nyerere had built a "man-eats-nothing" society. But Kenyatta could not bring himself to rise above ethnic loyalty. He favored his own tribe, the Kikuyu, massively skewing public resources to the Kikuyu heartland. Like many African leaders, Kenyatta had not made adequate preparations for his own succession. Two of Kenyatta's henchmen, both Kikuyu, wanted the job, and each blocked the other. In a sea of confusion they decided to appoint someone so hopeless that they could rule by proxy: they chose a poodle from a minority tribe. Step forward onto the world stage President Daniel arap Moi. In one key respect Moi was considerably less hopeless than the kingmakers had anticipated: he swiftly marginalized both them and the Kikuyu selectariat. Everything was reversed except for one constant: massive favoritism toward the president's own tribe, the Kalenjin.

As it happens, the Kalenjin tribe is itself a nice demonstration of how identity can be constructed. You might imagine that African tribes go right back to the primordial times of the birth of man. In fact, the Kalenjin go back all the way to 1942. With the Second World War being fought out in North Africa, the British wanted recruits for the Kings African Rifles and, sensibly enough, targeted their recruitment toward a large low-income area. The cheapest means of recruitment was to use the radio, but the area covered a wide range of dialects. Choosing one of the dialects in the middle of the range, each radio broadcast opened with the attention-grabbing phrase "I tell you, I tell you," not, of course in English, but in the dialect: "Kalenjin, kalenjin." In the appalling aftermath of the 2007 Kenyan elections, the Kalenjin led the violence. The tribe is the product of a radio program. Such is the stuff of ethnic identity.

While both Kenyatta and Moi favored their own tribes, neither devoted any serious priority to building a sense of national identity. There was no attempt to create a national language, and in the school system the history of each locality was given precedence over national history. Politically, the colonial system of chiefly power was largely left in place: the local big man became all important. As to interethnic equity, forget it. And despite its greater wealth, Kenya made no effort to build national symbols such as Dodoma.

The Kenyan elections of December 2007 provided an opportunity for a new set of politicians to fan the flames of a fire that had been lit by their predecessors. By far the main culprit was the opposition leader, Raila Odinga. Recall that the incumbent has the advantage in respect of bribery and miscounting, so the opposition is indeed more likely to resort to the cheaper strategy of playing on ethnic identity. Odinga ran a campaign that was tantamount to promising ethnic cleansing. His strategy was electorally successful because the Kikuyu, whom he targeted, constituted less than a quarter of the population. Odinga probably won the most votes. That he lost the election is probably due to ballot fraud. But if so, he was cheated out of a victory that was won by a strategy that in a proper democracy would have been illegal.

The difference in post-independence political strategies between Tanzania and Kenya was sufficiently stark to lay the foundations for a natural experiment: an attempt to build a sense of national identity, versus an attempt to reinforce tribal identity. However, a natural experiment needs much more than divergent strategies: the two places need to be otherwise comparable. The two countries were indeed pretty similar and certainly ethnically diverse: Kenya had forty-eight tribes, Tanzania even more. Miguel enhanced these country-level similarities by focusing on two districts, one Kenyan, the other Tanzanian. He selected them because they were even more similar than the countries themselves: Busia in Kenya, and Meatu in Tanzania. The international border, established in colonial times, had basically driven an arbitrary straight line through what until then had been one area. But divergent strategies and comparability are still not enough for a natural experiment. There needs to be some quantitatively measurable difference in outcomes: identity is a slippery sort of entity to observe. Miguel decided to measure the supply of some key public goods, such as the amount of money raised locally for schools, the provision of school facilities, and whether wells were in working order.

But if Busia was to be one observation and Meatu the other, there was not going to be any statistical power whatsoever: either Busia is going to be better than Meatu or it is going to be worse, and a priori, there is a 50 percent chance of finding either outcome. Miguel's key inspiration was to use the fact that both Busia and Meatu were composed of many localities. Some of these localities had high degrees of ethnic diversity whereas others were homogenous. He realized that he could use these differences in the degree of diversity between different localities *within* Busia and Meatu to see how much damage diversity was doing in each society.

In Busia, the Kenyan district, he found exactly the pattern that researchers have usually found when they investigate the consequences of ethnic diversity. The more diverse localities within Busia had worse public-goods provision than the more homogenous localities. What is more, the effect was really big. The average, fairly diverse locality had 25 percent less school funding per pupil than the homogenous localities. This was a problem fully recognized by head teachers in the ethnically diverse schools: they blamed ethnic rivalries for the unwillingness of parents to support the school.

How about Meatu, the Tanzanian district? The key test in the research design was whether ethnic diversity was similarly damaging there. There was just as much variation between localities in Meatu as in Busia: some localities were highly diverse and others were homogenous. It turned out not to matter at all: diversity had no discernible effect on public-goods provision. The statistics were supported by the interviews: Miguel received comments such as "We're all Tanzanians" and "This is Tanzania, we do not have that sort of problem here."

I hope I have given you a flavor of Miguel's study: it was, in fact, a beautifully crafted piece of social science. It is important because it provides pretty convincing evidence that Nyerere's strategy of building national identity had actually worked. Over a period of forty years, between independence and the survey on which these results are based, the damage normally caused by ethnic diversity had been dramatically reduced and perhaps even eliminated. Nyerere had turned a new country into a new nation.

Nyerere and Sukarno showed what could be done by leadership. Unfortunately, their approach was rare in the societies of the bottom billion. Far more common was that of Kenyatta and Moi in Kenya, where the consequences of a strategy of emphasizing ethnic identity over Kenyan identity are now all too apparent. As I write this I am trying to follow events in the aftermath of Kenya's election. Around one thousand Kenyans have died in ethnic violence. It is hard to discuss research in such a context. But recall that in Nigeria Pedro Vicente and I had conducted surveys during the presidential elections of April 2007. Since they had proved feasible, I decided to try the same approach during the Kenyan elections, which I anticipated would be rough. I put together a team. As you have seen, ethnic diversity in teams can be a source of strength: ours had a Kenyan, an American, a Belgian, a Mexican, and a German. This work was so recent that I can report only a few preliminary results.

The survey was conducted prior to the explosion of violence that followed the election. But even at this stage, five in every six Kenyans feared becoming victims of political violence, and one in ten had already been threatened about the consequences of voting the wrong way. Just as in Nigeria, electoral violence looks to have been a strategy of the weak: it was the government supporters who were the most fearful, and events proved them right. But the threats did not well upward from community-based antagonisms. The incitement to violence was seen as coming down from the organizations of the political parties. Violence against the Kikuyu was a deliberate electoral strategy of Raila Odinga.

Consistent with the allegations of fraud that followed the government declaration of victory, we found that at the time of our survey, which was a few days prior to the elections, the opposition was poised to win. Nor would this have come as a surprise to the Kenyan electorate: when asked how free and fair they expected the elections to be, 70 percent expected problems, and these fears were disproportionately high among opposition supporters. Ethnicity was all: only half of voters regarded their primary identity as being Kenyan. More revealingly, voting intentions were massively skewed by ethnicity. Not only did the Kikuyu vote for Kibaki and the Luo vote for Odinga, but even the tribes other than those of the candidates largely voted as ethnic bloc votes.

But here are the results that I think toll the death knell for ethnic politics. They concern the discipline that electoral competition is supposed to provide on government economic policies. In the years leading up to the election the Kenyan economy had been doing rather well: its fastest growth for more than two decades. Nor had the benefits of growth been confined to the Kikuyu. Even the Luo recognized that they had become better off. Kibaki even managed to get amazingly strong approval ratings from Luo respondents. It didn't help him. He was the wrong tribe and they were not going to vote for him: 98 percent of the Luo voted for Odinga. With this sort of voting behavior, there is little incentive for a president to provide national public goods: he might as well favor his own. The strong ethnic identities that Kenyan political leaders had fostered had effectively deprived electoral competition of its potential for holding a government to account. As for the other supposed benefit of elections, legitimacy, here is another comment from Koki Muli, the head of Kenya's Institute for Education in Democracy: "Do these people not care about legitimacy?"*

---

* "Kabaki Win Spurs Kenya Turmoil," *Financial Times*, December 31, 2007, p. 6.

Chapter 3

INSIDE THE CAULDRON:
POST-CONFLICT SETTLEMENTS

WITH THE MILLENNIUM CAME PEACE. The international community finally started to pay serious attention to the running sores of long-lasting civil wars. Peace conferences were called, pressure was put on the various sides, and a whole series of peace settlements achieved: Sri Lanka, Burundi, Southern Sudan, Sierra Leone, Angola, the Democratic Republic of the Congo, Bosnia, and Kosovo. While this was a splendid achievement, post-conflict situations are fragile; in the past around 40 percent of them have reverted to violence within a decade. In total these reversions account for around half of all the world's civil wars. So maintaining the post-conflict peace more effectively than in the past would be the single most effective way of reducing civil war. Is democracy the key to peace in these societies? International approaches to post-conflict situations are still in their infancy: a new organization, the Peace-Building Commission of the United Nations, is just finding its feet. The recent record is not entirely encouraging: here are a few examples.

Take the transitional government of the Democratic Republic of the Congo. Knowing that they had only three years in power before facing elections and the possible loss of office, ministers set about plundering the public purse. But the public purse was pretty small because tax revenue had withered away: as you will see, low taxation is part of the strategy of misgovernance. But plunder can extend beyond tax revenue. One strategy would be to borrow: saddle future citizens with liabilities and run off with the proceeds. Unfortunately for the new leaders of the Democratic Republic of the Congo, this strategy was not feasible: President Mobutu had already used it to the hilt so that the country was beyond its neck in debt. No bank was going to lend.

But there was an alternative. The Congo is mineral-rich. Much of these resources are unexploited because under President Mobutu it would have been folly for a company to incur the investment necessary to sink a mine. The president was stuck in what economists call the time-consistency problem: because he could not bind himself from confiscating investments, no sane company would make them. But by the time of the transitional government the global boom in commodity prices had changed the calculus of risk: it was worth paying a little something for the exploitation rights that the transitional government could legally confer. And so the ministers of the transitional government of the Democratic Republic of the Congo mortgaged the future of its citizens as surely as if they had issued debt, by selling off national assets at bargain prices. A few months ago I had lunch with one of the shrewd purchasers of these rights: a good lunch it was too. He became a little upset when I told him that the rights ought to be renegotiated.

Now take the most remarkable of all the conflict settlements, the peace in Southern Sudan achieved after many years of violence. The new government of Southern Sudan inherited an economic landscape that was virtually lunar: no provision whatsoever of public goods. No roads, no schools, no health care: nothing, not even buildings. The only public good was the security force, the Sudanese People's Liberation Army (SPLA), and that had just become redundant. There were, however, massive financial resources. Southern Sudan was sitting atop a newly opened oil field that straddled the border with Northern Sudan, and its share of the oil revenues provided an instant flow of $1.3 billion per year. On top of the oil revenues there was a huge aid inflow: quite appropriately, every agency wanted to help.

This was an environment par excellence in which priorities and sequence mattered. After all those years of sacrifice in the cause of liberation, the inhabitants of Southern Sudan might reasonably have hoped that their government would think through the critical path of building an effective state and get on with implementing it. So, two years on, what has happened? As a senior minister put it to me, "We've lost it." The most serious error was to devolve the power of public spending to the commanders of the military units that made up the SPLA. What did they do with their power? They expanded their own fighting units, putting their soldiers on the new public payroll. This alone has exhausted the oil revenue. And, of course, the government is now stuck. It can only free up the budget for productive uses by dismissing fighters who have just got themselves onto the gravy train. What else have ministers done: what about the aid money? Ministers themselves have decided not to bother living in Southern Sudan: they live in Nairobi where the public goods are better. They commute into the country they are responsible for governing because the donor agencies insist on holding meetings there. So what do the meetings reveal about ministerial priorities? Priority number one is for large, imposing ministerial headquarters buildings: you can surely picture the designs for the soaring concrete structures that will be the ministries of this and that.

I tend to think that governments get the private sector they deserve. In Southern Sudan there is one huge private investment. It is a five-star luxury hotel, sitting, like a space hotel, in the middle of nowhere. Because of the absence of public goods, there isn't even a road that leads to it. Who are the intended clients for the hotel?

Well, you will appreciate that Southern Sudan is not yet a major tourist destination, but it is a major destination for aid workers: that's the market. To entertain them, an international shopping mall has been built alongside. The aid agencies themselves have meanwhile spent their time squabbling over which agency should control the money: every agency wants to coordinate, and none wants to be coordinated. Currently the government of Southern Sudan is not sovereign: it shares sovereignty with the federal government of Sudan. But in 2011 there will be a referendum on full independence. Prepare to welcome the new country of Southern Sudan onto the world stage.

Now take Burundi, another long civil war recently settled. According to the terms of the settlement imposed by the international community, the peace was rapidly followed by an election. The most extreme among the various Hutu rebel movements won. Its early acts included imprisoning and torturing its opponents, embezzlement of the public purse for the purposes of importing guns for a private militia, and expulsion of United Nations peacekeeping troops. There was nothing the United Nations could do except to organize its withdrawal.

Now take Eritrea. Eritrea started its post-conflict independence from Ethiopia with the sort of rave international ratings of which other African governments could only dream. According to one investment rating Eritrea was going to be Africa's Singapore. Within a decade it was back at war with Ethiopia, followed by a coup by the president against his own government, half of the ministers of which were jailed. Military spending remains on a war footing with mass conscription. As I write, Eritrea has just expelled the peacekeepers guarding the buffer zone, not an encouraging step.

And finally the post-conflict darling: East Timor. This is the heroic little place that gained self-determination from Indonesia after a thirty-six-year struggle during which, due to the folly of President Sukarno's successor, President Suharto, it was turned into a colony instead of welcomed as part of the nation. It joined the ranks of the international community to a chorus of congratulation. It is perhaps ungracious to point out that if every group of eight hundred thousand people was granted the right to self-determination, the world would have around eight thousand countries. In other words, independence for East Timor did not pass the test proposed by the moral philosopher Immanuel Kant: "What if everybody did that?"

But, never mind; how has heroic little East Timor progressed since its independence in 2001? East Timor was one of the 40 percent of post-conflict situations that did not make it through the first decade without a reversion to violence. In 2006 one of its leading politicians was found to be importing arms for his private militia. A large disaffected group in the army that came from the western part of East Timor attempted a coup and then retreated into the mountains: sure enough, the same mountains where the civil war had been fought. In the ensuing struggle a tenth of the population was displaced. Had not two thousand Australian troops promptly arrived to put down the coup, a prolonged civil war might have led to the entry onto the world stage of a new sovereign country of West-East-Timor.

DESPITE ITS IMPORTANCE, UNTIL RECENTLY I had shied off trying to investigate what determines whether a post-conflict peace endures. In statistical terms it is a difficult question because of the relatively small number of pertinent observations. By 2006 we had accumulated data on sixty-six countries, which was at last sufficient to be worth investigating. This time my team was Anke Hoeffler and Mans Söderbom, a very smart Swede. We decided to cast our net wide and investigate on an equal footing all the possible influences on the duration of peace: political, social, economic, and military.

Let's start with where we left off: democracy and elections. The standard approach of the international community to the end of a civil war is to insist on a democratic constitution and crown this after a few years by an election. This is the theory of legitimacy and accountability at its clearest. Peace will be secured by an election because the winner will be recognized as legitimate by the population, making violent opposition more difficult. Not only will the elected government be recognized as legitimate, the democratic process will ensure that it will need to be inclusive and so there will be less reason for grievance: the government will be accountable to its citizens. It is time to look at the evidence.

We first checked whether the type of polity affected whether a post-conflict country reverted to violence. Again we used the twenty-one-point scale of the Polity IV index, searching along it to see whether any part of the range was significantly safer than any other. We did not like what we found. There was a portion of the range that was significantly safer, but it was the range of intense autocracy: between −10 and −5. For the countries in this range the risk of reversion to conflict was much lower: not 40 percent, which was the overall average, but around 25 percent. Correspondingly, the polities that were less repressive, that is, with a score of −4 or better, had an above average risk of reversion to conflict: not 40 percent but an astounding 70 percent.

To think concretely, and to take examples that occurred sufficiently recently not to be driving the results, in the early years of the new millennium both Angola and Sri Lanka made it to peace. Angola continued to be one of the most repressive regimes on earth, whereas Sri Lanka was a long-established democracy. The peace in Angola has held firm, and I expect that it will continue to do so. The peace in Sri Lanka has already fallen apart: rich-country governments have heaped the lion's share of the blame on the Sri Lankan government rather than on the Tigers, just as they tended to blame the government of Colombia for the resumption of the war against the FARC, and the government of Uganda for the running war against the Lord's Resistance Army. I am ready to admit that all three of these governments have probably made mistakes, but what is manifest is that all three of them are saintly when compared with that of Angola. In other words, a more democratic polity does not necessarily make peace more likely.

So much for the effect of the polity. We pressed on to the effect of elections, introducing them into our model of risks during the post-conflict decade. There were plenty of elections, but at first we could not make sense of them: there seemed to be no clear effect at all. Surely, in the highly charged environment that is typical of post-conflict, the key political event of an election could not wash over the society leaving no significant effect. And then we hit on it. A post-conflict election *shifts* the risk of conflict reversion. In the year before the election the risk of going back to violence is very sharply reduced: the society looks to have reached safety. But in the year after the election the risk explodes upward. The net effect of the election is to make the society *more dangerous*.

Why do post-conflict elections have this effect? Well, at this point we have to leap off the statistical results and start to speculate. Here is my guess. In the run-up to an election there is a strong incentive for the parties to participate: after all, this is the route to power. So energies get diverted into campaigning and so risks fall. But then comes the election result. Someone has won, and someone has lost. Of course, if this was a genuine democracy the winning party would say the sort of things that winning parties usually say in genuine democracies: we will govern on behalf of all the people. And because of checks and balances that constrained its power while in office, it would more or less have to do so. If it was a genuine democracy the losing party would say the sort of things that losing parties usually say in genuine democracies: we congratulate the winner and will be a loyal opposition. Because of the restraints on abuse of power, the losing party would know that it still had a good chance of attaining power within five years. Post-conflict situations are not usually like this. The winner gleefully anticipates untrammeled power: no checks and balances here. The loser anticipates its fate under the thumb of its opponents and knows there is but one recourse: back to violence.

Let's go back to the first post-conflict situation I described: the Democratic Republic of the Congo. President Mobutu had been ousted by the rebel leader Laurent-Désiré Kabila, with the military backing of Rwanda and Uganda. In 2001 Kabila had been assassinated and his young son Joseph inherited the throne, indeed becoming the youngest head of state in the world. I apologize for that slip: "throne" might give the entirely wrong impression that the true purpose of the ostensibly Maoist rebellion had been to install an absolute monarchy. Let me at once correct that impression: the young Joseph was appointed as the next president by due constitutional process. Indeed, since the international community held the key cards in this situation, they called the shots, other than the one that got Laurent-Désiré. Recall that the government was up to its eyes in debt, was chronically short of revenue, and lacked an effective army. So President Kabila II had little choice but to acquiesce in holding a post-conflict election.

The election was to be in two rounds, somewhat like that in France, with the second and decisive round set for October 29, 2006. The international community was sufficiently confident of the legitimacy and accountability model that it set the date for the withdrawal of its peacekeeping forces as October 30, 2006. This was the denial of reality at its most absurd: democrazy in action. If our results were right, in one sense the strategy of the international community was understandable. If events in the Democratic Republic of the Congo ran to form, the year before the election would be remarkably peaceful, creating the impression that the society was now over the period of high risk. Since international peacekeeping is both enormously expensive and highly unpopular with electorates in the high-income countries that provide the troops, there is strong pressure to "bring the boys home" as soon as there looks to be no further need for them. So it is not surprising that the post-conflict election should be used as the milestone, in the surreal technical jargon of peacekeeping, for troop withdrawal. Or, in the more familiar sound bite, elections are the exit strategy. How this strategy played out in the Democratic Republic of the Congo I will return to shortly.

If you think about it, our results suggest that a post-conflict election is inappropriate as a milestone: it is more like a tombstone. Of course, it depends whether peacekeeping works: if it doesn't work then the boys might as well be brought home and any sort of stone will do. So it is time to turn from elections to peacekeeping.

We asked the United Nations for data on its peacekeeping operations. The good news was that they had pretty complete records. Unfortunately, the records were not organized for quantitative analysis: it took our research assistants seven months to put them into shape. But eventually we had information, country by country and year by year, on the numbers of troops and their cost. It was time to see whether peacekeepers helped to keep the peace. Somewhat to our surprise we got clear results: peacekeeping seems to work. Expenditure on peacekeeping strongly and significantly reduces the risk that a post-conflict situation will revert to civil war.

By now you will realize that the standard concern is whether such results are spurious because of reverse causality. For example, if the troops are systematically sent only to the safer post-conflict countries, they will appear to be successful in keeping the peace but the result will not be causal. And so we tried to find something that would explain the allocation of peacekeeping troops but that was otherwise unrelated to the risk of conflict reversion. Whatever we tried, we were unable to get a good explanation for the allocation of troops, and so we turned to the academic literature. Nicolas Sambanis, a young Greek political scientist whom I had once worked with, had just coauthored a book on post-conflict peacekeeping with Michael Doyle, who used to be head of research at the United Nations and is a world authority on peacekeeping. They had concluded that the political decision process that assigned troops to post-conflict situations was so complex as to defy being modeled. The various members of the Security Council who took the decision were involved in such byzantine horse-trading that any particular decision was close to being random. This explained why we were unable to find good predictors and also suggested that we were not facing a severe problem of reverse causality.

Nevertheless, we were able to make one helpful check. The decision as to how many troops to send into a post-conflict situation can conceptually be split into two stages: first, should troops be sent at all, and, then, if it is decided to send troops, how many should be sent? We realized that we could learn something about the motivations for sending troops by looking at that first decision: should they be sent at all? We found that the decision to send troops at all was associated with a significantly higher risk of reversion to violence. The most plausible way of interpreting this is that troops tend to be sent to places that are more at risk. We cannot tell whether the same is true of the decision as to how many troops to send. We just know that given the decision to commit troops, the more that were sent, the safer was the society. If, in fact, the number of troops sent is motivated by the same concerns as seem to motivate the prior decision of whether to send any, then they are being sent in the greatest numbers to the most dangerous places. How would this affect our results, which implicitly assume that they are assigned randomly? Its effect would be that our results would *understate* the true effectiveness of peacekeepers. The truth would be that places with many peacekeepers have a lower rate of reversion to conflict *despite* intrinsically being more at risk. So our assumption that their numbers are unrelated to intrinsic risk may well be conservative.

I had the results on post-conflict elections and on the efficacy of peacekeeping by the summer of 2006 and shared them with the appropriate parts of the international community. I was particularly concerned that the proposed strategy in the Democratic Republic of the Congo of troop withdrawals the day after the election, which was due to be implemented within a couple of months, looked unwise. I was promptly invited to address the new Peace-Building Commission of the United Nations, and also shared the results with the French government, who were supplying the largest component of the peace-keeping troops. I learned that the military commanders were themselves highly skeptical of the plan for troop withdrawal. In the event, the aftermath of the election became so violently unstable so rapidly that instead of troops being flown out, they had to be flown in. Within a few months there was a shoot-out between the private army of Bemba, who lost the election, and the government army of Kabila II, the incumbent winner. Bemba's forces lost the shoot-out, and he himself sought protection in an embassy before fleeing to Europe, where he is now in exile. His exit has not restored order: the Democratic Republic of the Congo continues to be dangerous.

Even if international peacekeeping is effective it faces problems. It is expensive and unpopular. Some of the post-conflict governments get indignant about the intrusion: the Department of Peace-keeping Operations of the United Nations (DPKO) has become the new International Monetary Fund, a challenge to the unrestrained sovereignty of governments keen on asserting their power. It is also understandably unpopular with the electorates of the countries that supply the troops: no one wants his son or daughter to be exposed to the risks of peacekeeping.

Is there an alternative? I could think of two other possibilities. The first is what is known as over-the-horizon guarantees. It is what the British government is doing in Sierra Leone. For the past few years there have been only eighty British troops stationed in the country, but the government has been given a ten-year undertaking that if there is trouble, the troops will be flown in overnight. Perhaps this has helped stabilize the society. Sierra Leone is, at least in terms of reversion to violence, a major success. It has even weathered post-conflict elections and a change of government. The problem with the Sierra Leone example is that it is just that: one example. You cannot perform a statistical analysis on one observation, and so there is no way of knowing whether in general such guarantees would be effective. Or is there?

I started to think whether there was anything analogous in the past to what the British are now doing in Sierra Leone. Sure enough, the French had provided security guarantees to their client countries in Africa for years. In fact, with the typical logic of international coordination, they had abandoned it only just before the British started to do it. The French security guarantees were informal, but they were most surely for real. They were backed by a series of French military bases across Francophone West Africa. They had started with independence and rolled on until the French government got caught up trying to implement its informal guarantee defending the Hutu regime in Rwanda in 1994. If you remember, there were French troops stationed in Rwanda as the Tutsi rebel forces invaded from Uganda and as the Hutu regime set about its genocide. The French came disturbingly close to finding themselves propping up a regime implementing genocide and only just pulled back in time. After that President Chirac ordered a rethink and announced a new policy toward Africa: military intervention began to look anachronistic. The first test of this new policy was the coup d'état in Cote d'Ivoire in 1999. The French old guard wanted to intervene to put it down, but President Chirac vetoed intervention. So we can date the credible prospect of French intervention from independence until the mid-1990s. After their military catastrophe of the battle of Dien Bien Phu in Vietnam the French were in no position to extend their military guarantee across the whole of the Francophone world; it was basically credible only in West and Central Africa, and it had lasted for around thirty years. This was, however, a large enough group of countries, for a sufficiently long period, to be amenable to statistical analysis.

The key question was whether this guarantee had actually reduced the incidence of civil war. This question needs a model of the
risk of civil war. Such a model can be used to address a range of important questions, but here I will just give you this particular answer.
Did the French informal security guarantee reduce the incidence of
civil war? We found that it was highly effective. Francophone Africa had characteristics that would otherwise have made it prone to
warfare: the actual incidence was much lower than would have been
expected. Statistically, the guarantee significantly and substantially
reduced the risk of conflict by nearly three-quarters.

But was the military guarantee the reason for this remarkable
reduction in conflict? Could it have been something else associated
with the French presence? For example, in response to French opposition to the invasion of Iraq, some Americans accused the French
of an excessive aversion to force: what was that ill-judged phrase,
"cheese-eating surrender monkeys"? Perhaps French culture had
inculcated pacific values? While this might seem implausible to
anyone aware of French military history, we decided that no stone
should be left unturned. If the reduction in the risk of conflict was
due to culture rather than the security guarantee, it would reach the
parts of La Francophonie that the guarantee did not cover. It didn't:
the enhanced security was unique to West and Central Africa, the
region covered by the French military bases. To my mind this is reasonably convincing. Over-the-horizon guarantees look as though
they work. As I was finishing this book Chad exploded into civil
war: rebels reached the gates of the presidential palace. As the crisis
unfolded the French position rapidly shifted. Initially the French
announced that they had no intention of intervening militarily.
Within a week they had thought better of it and issued a security
guarantee: the rebels would be repelled by French force unless they
withdrew. The French had a large military base right there in Chad:
the rebels withdrew.

It is time to move on from the politics and the military. What else drives post-conflict risks? Surely the economy enters some-where? In fact it enters twice over. The lower the income, the higher the risk of conflict reversion; and the slower the economic recovery, the higher the risk. Both of these have implications. If low-income countries face higher risks of conflict reversion, other things equal, the international community should be allocating peacekeeping troops disproportionately to those post-conflict situations with the lowest income. This would indeed provide a useful rule-of-thumb to cut through all the horse-trading that Doyle and Sambanis found to be dominating the decision on the Security Council. A further im-plication is that, again other things equal, strategies that enhance the economic recovery are going to be peace-enhancing: raising growth and cumulatively augmenting the level of income.

So how to rebuild a shattered post-conflict economy? The problem with economic interventions is that they are not exactly the cavalry. It is possible to destroy an economy quite rapidly, as Presi-dent Mugabe has convincingly demonstrated, but putting Humpty together again takes time. If average income can grow at 7 percent a year, which is entirely possible in post-conflict situations, then the level of income doubles in a decade, and so by the end of the decade risks are substantially lower. But this is the time frame for economic recovery, not two or three years.

The story so far is that the post-conflict decade is dangerous and that there seems to be no clear political quick fix. In particular, elections and democracy, at least in the form found in the typical post-conflict situation, do not bring risks down. Economic recovery works but it takes a long time. The one thing that seems to work quickly is international peacekeeping, but it is politically difficult to sustain peacekeeping for the length of time needed for the economy to recover. Is prolonged peacekeeping necessary, even if only in the form of over-the-horizon guarantees? There is one remaining pos-sibility. Perhaps the key risks occur early in the decade and are fol-lowed by a safe period. In that case peacekeeping could be brief.

That would make it politically much easier. Since an option that is politically easy is far more likely to be adopted, it was worth investigating. The risk of going back to conflict does seem gradually to decline with time, but don't hold your breath. Time heals, but its effects seem to be decade by decade rather than year by year. The first four years after the end of a conflict are perhaps somewhat more dangerous than the next six, but the effect is not statistically significant. Within the post-conflict decade there is no safe period.

So where does this leave us? Economic recovery is to my mind the only genuine exit strategy for peacekeeping. I think we need to dismiss the illusion that elections are the milestone and face the long haul of building the economy. It may well not be necessary to keep many peacekeeping troops on the ground throughout the decade: an initial military presence may well be able to evolve successfully into an over-the-horizon guarantee. But any such guarantee must be credible: the French guarantee was made credible by its military bases, and the British guarantee was credible because during the conflict they indeed flew in overnight to check the Revolutionary United Front (RUF) forces that were set on occupying the capital, Freetown. The British forces held off the RUF on the outskirts of the capital at little place called Waterloo. But they only arrived just in the nick of time: as Lord Wellington said of the former, somewhat grander battle, it was "a damned close thing."

So, if economic recovery is the exit strategy, how can it be facilitated? What policies work, and can donors help? Anke and I had already done a little work on the payoff to post-conflict aid: we found that it was significantly more effective than aid at other times. This is not surprising: post-conflict recovery was the initial rationale for the international aid agencies. But I decided it would be worth looking more closely at what could be done to revive the economy. For this work I teamed up with Victor Davies, a doctoral student from the classic post-conflict country of Sierra Leone, and Chris Adam, a close colleague at Oxford. I also worked with Marguerite Duponchel, a doctoral student at the Sorbonne, where I teach as a visitor. Although I will try to make what you are about to encounter come across as a seamless web of research, it was not like that at the time.

Some important uses of aid for post-conflict are blindingly obvious: it pays for the reconstruction of infrastructure. But here is one that is much less obvious: countering inflation. High inflation is a pretty disastrous macroeconomic strategy: essentially a policy of desperation. Normally governments keep inflation moderate. This is despite the fact that in the short run they could fleece the economy by printing money. Inflation is a tax that most people do not recognize as a tax. Governments restrain themselves because of the alarming dynamics that lead to hyperinflation. The only governments that resort to it are therefore governments that are desperate: ones that have almost given up on the future because their struggle to survive in the present is all-consuming. One likely reason that President Mugabe has given up on the future is that he is himself eighty-four years old: the oldest head of government in the world. Ironically, Zimbabwean citizens are on average among the youngest in the world, so you might otherwise have expected this society to be extremely conscious of the future. Unfortunately, even ordinary citizens have good reason to discount the future: life expectancy in Zimbabwe is rock-bottom.

Zimbabwe is truly rare in being a peacetime society in full-blooded hyperinflation. But we wondered whether during civil war governments might often get desperate. Normal tax revenues decline as the economy contracts in the face of violence. These declining revenues collide with exploding expenditures as the military demands a larger budget: during a civil war military spending typically nearly doubles. So, we reasoned, governments were likely to resort to the printing press. This supposition turned out to be correct, but our concern was not what happened to the economy during civil war but its implications for the post-conflict recovery. The legacy of high inflation is that people expect inflation to continue and learn to economize on holding cash. If the government wants to get back to where it started—low inflation and people sufficiently confident to hold the currency—it will need a prolonged period of fiscal restraint. In effect, during the civil war the government's inflation strategy has been equivalent to borrowing: it has created the liability of expected inflation. The post-conflict government now faces the need to wipe out this liability. But it faces its own desperate fiscal challenges.

Tax revenues will take time to rebuild. Typically the economy will have become less formal as firms try to escape taxation. Raising taxes too rapidly retards the process of rebuilding formal activity. In Sierra Leone the head of the chamber of commerce told me how his membership had dwindled as firms had gone informal. Yet there are huge demands for government spending: rebuilding infrastructure, alleviating crises of collapsed social provision, creating jobs for unemployed youth. We found that an unnoticed benefit of post-conflict aid is that it helps to square this circle. The government no longer needs to use the inflation option so aggressively, and this enables confidence in the currency to return. Like other aspects of economic recovery, this is a slow process: this use of aid is an investment in confidence and it takes more than a decade before citizens are back to their pre-conflict willingness to hold the currency. But without post-conflict aid it takes far longer.

Does it matter? Well, it turns out that inflation is particularly damaging in post-conflict economies for a simple reason. During a civil war residents shift their assets abroad: it is called capital flight. Such flight is a major phenomenon across the societies of the bottom billion. With Tara MacIndoe, a Zimbabwean graduate student, Anke and I have estimated the proportion of Africa's private wealth that is held outside the region. By 2004 it had reached the astound-

ing figure of 36 percent: more than a third of Africa's own wealth is
outside the region. Unsurprisingly, during and after civil war capital flight is even more severe than this average. For the post-conflict
period the legacy of accumulated capital flight is potentially a lifeline: if only the money can be attracted back. But usually instead of
being a lifeline it is a continuing hemorrhage: faced with a high risk
of conflict reversion, people continue to shift their money out. This
is a collective action problem: in aggregate, capital flight retards economic recovery and so makes conflict reversion more likely. Everyone has an interest that everyone else should keep his capital in the
country, but his individual interest is to shift it out.

So how does this relate to inflation? Victor discovered that capital flight in post-conflict situations was particularly sensitive to inflation, much more than in normal peacetime conditions. We are not
sure why: perhaps high inflation is seen as a sign of future instability,
indicating that the government is itself discounting the future heavily. But the implication is that aid during the post-conflict period is
particularly effective. By enabling the government to reduce inflation, aid is geared up by reduced capital flight, and indeed possibly
by capital repatriation.

It is not only capital that is lost during conflict, it is also skills.
With Marguerite I have tried to analyze the effect of the conflict in
Sierra Leone on the skill base of private firms. There are few data
available for post-conflict economies, and so this work was right on
the edge of what is possible. We were given a new survey of firms and
workers that had been conducted by the United Nations Development Program (UNDP). Although it included a lot of information
on worker training, by itself it was largely uninformative. We needed
to match it with other information. A team of researchers at Berkeley had surveyed households across Sierra Leone, recording the family history of deaths through the violence of the civil war. The team
kindly shared their data with us, and this allowed us to build a picture
of how the incidence of violence differed, locality by locality.

Our idea was to see whether the violence had destroyed jobs and skills by matching this picture of the incidence of violence against the survey of firms and workers. There was, however, a further problem: we needed to know what the economy had been like locality by locality before the violence. I sent Marguerite off to the libraries to search through the archives: one of the useful features of Oxford is that it has good collections of musty documents, and I guessed that there must at some time in the past have been a survey of firms in Sierra Leone. Sure enough, after a lot of searching Marguerite discovered that there had indeed been such a survey, but Oxford had only a partial set of the report, done some thirty-seven years ago. Our library ran a national search, and the missing volumes were eventually found. I suppose that in a few years we will be able to download all these old documents from the Web and such searches will become a thing of the past.

At last we thought we were in business: how had the violence affected firms, jobs, and skills? And then we started to worry: suppose that the violence had been targeted on the poorest places; superficially it would look as though the violence had impoverished them, but it would be spurious. We were in danger of getting causality the wrong way around. There is a way around such problems: you need to find something that increases the risk of violence but that does not directly affect the economy. Fortunately, we hit on one: the RUF rebel force had been based in Liberia: one lawless state had served as a safe haven for spewing violence into its neighbor. And so within each district of Sierra Leone differences between localities in the distance from Liberia turned out to be a pretty good predictor of the scale of the violence; equally important, other than through violence it did not much affect the economy. We were at last truly in business. It took around three months to reach that point and about three days to move from the data to the results. Until that point you cannot tell whether all that effort has been wasted. It was, of course, Marguerite who was facing the risk: had I marched her up a blind alley, or was she going to find results that could be defended in a doctorate?

What we found looks interesting. Seven years after the end of the conflict there was no trace of any effect of the violence on either the number of firms or the number of jobs. By looking at the size of firms and the year when they were established, we found that where violence had been intense, firms had shrunk. Although to an extent these firms had bounced back once the fighting was over, it had left a significant legacy: it had sharply reduced worker productivity. Responding to the problem of low productivity, firms in the previously violent districts were more likely to be undertaking basic training of their workers. Evidently, violence had deskilled the workforce. The overall picture was of a flexible private economy that had been ravaged: firms reestablished and workers could find jobs at some pitiful wage, but the skills that would have justified higher wages had been destroyed. More than forty years ago Nobel laureate Ken Arrow had the key insight into the process of skill accumulation in a society. He called it "learning by doing": productivity rises in an activity with practice. Conceptually, there is a counterpart to learning by doing: "forgetting by not doing." In the prosperous economic context of Arrow's work, this logical possibility was not even sufficiently important to warrant a footnote. But for the economics of civil war it matters. Civil war is development in reverse, and Arrow's model is running backward.

Are there any forgotten skills that are particularly important for the post-conflict recovery? I think that there are. In fact, they are too mundane for the aid agencies to notice them. The agencies spend a fortune on the rather ambitious agenda of inculcating attitudes of reconciliation, which recent studies find to be ineffective. But they neglect the obvious. During civil war the sector that collapses most severely is construction: the society is hell-bent on destruction and so nobody is investing in buildings and infrastructure. The construction sector uses a lot of unskilled labor, which has the potential for job creation among the unemployed youth that constitute a post-conflict powder keg. But even the construction sector needs skills: you cannot build a wall with only unskilled labor. And so the fact that the basic skills needed for construction have atrophied during the war becomes an impediment to rapid economic recovery. Showing Marguerite the half-collapsed roof of a building, her guide explained, "My grandfather would have known how to repair that, but I don't."

As donors and the government try to rebuild, their spending pushes up construction prices and so gets dissipated: the skills shortage is a bottleneck in reconstruction. For example, in Liberia the cost of constructing a school has roughly doubled. The construction bottlenecks need to be broken. Some donors do this by hiring Chinese contractors: the Chinese face no bottlenecks because they routinely bring in absolutely everything, including the entire workforce. But resorting to the Chinese throws out the main short-term benefit from the recovery of the construction sector, which is to generate jobs for young men. This looks likely to be critical in bringing down the risks of conflict. So the solution to the skill bottleneck is training. Post-conflict situations need squads of bricklayers, plumbers, welders, and so forth, who set about training young men. Unfortunately, it is too mundane for the development agencies to organize it. We need Bricklayers Without Borders.

WHERE WE HAVE GOT TO is that post-conflict societies are fragile, and that there does not seem to be a simple political solution. The strategy that makes a difference, really bringing down risks, is peacekeeping, evolving into an over-the-horizon guarantee, for which the exit strategy is economic recovery speeded by aid.

Each of these components seems to make a significant difference, but this does not mean that they are necessarily worth doing: they might simply be too costly to be worth the candle. How can we judge whether an intervention is worthwhile? The answer provided by my profession is cost-benefit analysis, which is just a fancy term for weighing up the pros and cons. So let's weigh them up. Since any military intervention is now controversial, I am going to focus on peacekeeping. Is it cost-effective? Two yardsticks are useful: what is the ratio of costs to benefits, and what is the net gain?

The reduction in risk achieved by peacekeeping forces depends upon their scale of deployment. Our estimates should be seen as rough approximations: the precise results depend upon technical choices, and while we think our particular choices are defensible, they could surely be improved upon. We estimate that an annual expenditure of $100 million on peacekeepers reduces the cumulative ten-year risk of reversion to conflict very substantially from around 38 percent to 17 percent. If peacekeeping forces are scaled up, the risk falls further. At $200 million per year the risk is down to around 13 percent, and at $500 million it is down to 9 percent. The next step is to convert this reduction in risk into a benefit. For that we need an estimate of the costs of conflict. I am going to use the figure of $20 billion. Although this seems enormous, for the typical civil war in a country of the bottom billion it is at the lower bound of the range of realistic costs: it is likely to be considerably too conservative. If a civil war inflicts costs on the society of around $20 billion, then of course avoiding one confers benefits of $20 billion. By extension, a strategy that replaced an inevitable war with one that occurred only if a flipped coin came up heads would therefore be worth $10 billion.

More generally, each percentage point reduction in the risk of a civil war is worth around $200 million. Recall that peacekeeping at the level of $100 million per year sustained over the post-conflict decade reduces the risk of civil war by 21 percentage points. So the value of the benefit is around $4.2 billion. Since the peacekeepers are there for a decade, their total cost is $1 billion. We are at last ready for the punch-line number: the ratio of benefits to costs is better than four to one. Peacekeeping looks to be very good value. Given the difficulties of estimation, this is likely to be a long way out. Technically, it is possible to estimate statistical confidence intervals, and I have duly done so. But a much surer way of building confidence is the challenge of rival researchers. As other estimates build up we will learn the range of credible answers. I make no claim for the numbers that I have just presented, other than that they can be very wrong before peacekeeping starts to look a waste of money. I was invited to present these results, and the case for peacekeeping, to the panel of judges for the 2008 Copenhagen Consensus. The panel assesses ten rival research teams making the case for international public money to be spent on something. The process is terrifying: the panel consists of Nobel laureate economists, and defending my work before it reminded me of my doctoral viva more than thirty years earlier. The verdict of the panel was that peacekeeping was selected as one of the approved expenditures. In their words: "The expert panel found that peacekeeping forces in post-conflict societies could provide fair value for cost."

While the ratio of benefits to costs is a useful guide to action, it is not in fact the end of the story. As will be evident from these figures, peacekeeping forces appear to be subject to diminishing returns: as you keep on expanding the force, you get less bang for the buck. Scale is not everything, of course; quality matters as well. The large United Nations force originally assigned to Sierra Leone was useless because its soldiers lacked the mandate and the willingness to fight. But, given the level of quality, scale matters. Since the bang for the buck diminishes, there is at least potentially some ideal scale of intervention. Although the notion of an ideal scale might seem esoteric, it can be given a very clear meaning. The ideal scale is reached when a further increase in peacekeeping expenditure would generate additional benefits that just equal their cost: beyond that, expansion would be wasteful. In principle, the analysis that yields the ratio of costs to benefits also yields the ideal scale of peacekeeping operations. Obviously, peacekeeping has to be tailored to particular circumstances, and so the question is not answerable statistically on any realistic array of data.

Given the primitive nature of my own model it is definitely a bridge too far, but let me take it nevertheless to illustrate the approach. The point at which additional benefits match additional costs is reached when the peacekeeping force is somewhere between $100 million and $200 million. At $200 million the risk of conflict reversion is reduced by around 25 percentage points and so is worth around $5 billion. Its cost over the decade is $2 billion: effective peacekeeping is not cheap. While peacekeeping on this scale has a much less impressive ratio of benefits to costs, it still leads to a large overall gain of around $3 billion. These estimates are conservative because they are based on a cost of war that omits many important elements, and so the likely overall gain from a peacekeeping operation is, I suspect, even larger. The core role of politicians is to mobilize the collective action that supplies public goods with benefits far in excess of their costs. Peacekeeping is such a public good.

Is this quantification a complete fantasy? I think that in the Democratic Republic of the Congo there is a reasonable case that the presence of peacekeepers has averted a catastrophe. If so, it has surely been good value for money. Quantification forces you to flesh out such judgments. Nobody is going to be so foolish as to base policy on numbers. But it is surely useful, given that such huge amounts of money are being spent and that lives are on the line, to try to move beyond gut instinct. Indeed, peacekeeping is quite unpopular with aid agencies: they see large amounts of money being channeled through their ministries of defense and would like it diverted into their own budgets. Such decisions should not be taken on the basis of turf wars: in the end, the question of whether peacekeeping gives value for money has to be faced.

While the initial maintenance of peace cannot credibly be done without troops on the ground, the British experience to date in Sierra Leone suggests that it may be possible to phase out the bulk of international troops after, say, five years, replacing them with a guarantee made credible by a rapid-reaction force. The French security guarantee for Francophone Africa prior to the late 1990s reduced the risk of a civil war breaking out in the typical country of Francophone Africa from around 10 percent in any five-year period to around 3 percent. The risk reduction achieved by the French security guarantee can perhaps give some guidance as to whether over-the-horizon guarantees are cost-effective.

When I started to think about how to do a cost-benefit analysis of the French guarantee, I expected to need three components. One would be a figure for the reduction in the risk of conflict: I have just given you that, from 10 percent to 3 percent. The second would be the cost of achieving this reduction in risk. I asked the French Treasury for an estimate of approximately how much its rapid-reaction force had cost, and they gave me a ballpark figure. As a ballpark it must be treated with heavy caveats, but they thought that it might have been around $1 billion per year. This is equivalent to a super-force of peacekeepers in a single country, but this very scale presumably added to its credibility. Indeed, the guarantee force must evidently be at least as large as that needed in the largest envisaged operation. The third component in a cost-benefit estimate would normally be the cost of an averted conflict. For peacekeeping I used the figure of $20 billion, but for over-the-horizon guarantees, I hit on a way of avoiding the need for any figure at all. This was to value the over-the-horizon guarantee not relative to the absence of any peacekeeping but relative to the continued presence of peacekeepers in the country. This way I could pose the question as "How many peacekeepers could be brought home if a guarantee was put in place while leaving the risk of conflict unaltered?" An advantage was that posed in this way, I did not need to value any change in the risk of conflict: by design there would be no change.

The gain from the release of peacekeepers on the ground clearly depends upon the size of forces and on the number of situations covered by a single rapid-reaction force. For example, I estimated that if the initial forces were costing $500 million, then they could be scaled back to around $100 million. This estimate is purely illustrative. There are simply not enough instances of peacekeepers being partially withdrawn and replaced by an over-the-horizon guarantee for us to be able to make a defensible estimate. But the illustration is a guide to the judgments that will need to underpin actual decisions. For example, in my example a soldier based over-the-horizon in a rapid-reserve unit is much less effective than one based in the conflict situation, and this seems plausible. However, the rapid-reaction force might nevertheless be more cost-effective if the same force can cover more than one situation. A rapid-reaction force is analogous to a fire brigade, whereas a force in situ is analogous to a sprinkler system. In my example, a rapid-reaction force would be cost-effective if it could provide cover for three post-conflict situations. And this is before counting the benefit that for most of the time soldiers would not need to be away from home.

WHAT SHOULD WE CONCLUDE FROM all this? Post-conflict situations are fragile and there is usually no simple political fix. Peacekeeping, phasing into an over-the-horizon guarantee, looks to be the key instrument for post-conflict peace. The disquiet that it now evokes both in the countries that send troops and in the regions that receive them is understandable but misplaced. Even peacekeeping is not a quick fix: it needs to be sustained for around a decade. Peacekeeping is the handmaiden of economic recovery, not its rival, and so budget wars between the aid agencies and defense ministries of the high-income countries are misplaced. Building post-conflict peace is expensive, and both will need large budgets. The case for a large budget for peacekeeping is that it looks to be very good value. We should learn to support it. Post-conflict situations also need big aid. The aid-assisted economic recovery is the true exit strategy for the peacekeepers.

Part II

FACING REALITY:

NASTY, BRUTISH,

AND LONG

Chapter 4

GUNS: FUELING THE FIRE

A S W E A R E T O L D : G U N S don't kill people, people kill people. The genocide in Rwanda showed that mass killing does not need guns: the Hutu government managed to generate a slaughter of more than half a million people, largely with machetes. But guns become necessary if your opponents have them. The Hutu government could slaughter Tutsi civilians with machetes because they were unarmed, but rebels face government armies and so they need guns. No rebel guns, no rebellion, and so no nasty, brutish, and long civil war. And because the government next door has guns, our government needs guns: a government without guns cannot defend its citizens against a neighbor with guns. That is the message that many of our politicians thrive on: national security is the ultimate national public good and military spending is the way to achieve it.

Like many issues, whether the ready availability of guns makes a society more dangerous or less dangerous is an empirical matter. Despite the overheated political positions, there are three perfectly sensible possibilities. Cheap and plentiful guns may increase the risk of violence. Alternatively, they may make violence so dangerous that they deter it. Finally, guns may be plentiful where there is a lot of violence, but this may be because in societies that are violent for whatever underlying reason, people make sure they have guns around: the guns are a consequence, not a cause. Ideologues seem to think that such issues can be settled a priori: turning to the evidence challenges superficially plausible belief systems with the possibility that they are make-believe.

PROBABLY THE MOST IMPORTANT QUESTION to be asked about guns is whether they work as a deterrent. However, in order to answer this question, we first need to know why governments buy so many of them. Somehow we have to be able to sort out the chicken-and-egg causality problem: the risk of violence affects military spending, and military spending affects the risk of violence. If military spending keeps a society safe it is worth every cent. But before slipping into such easy thought patterns I decided to investigate what has actually been driving it. By this I do not mean answers such as "the military-industrial complex" about which the Marxists of my youth used to fantasize.

During the Cold War there used to be an academic industry that studied the arms race between NATO and the Warsaw Pact. But with the end of the Cold War, that collapsed, and nothing has replaced it. There were scarcely any recent studies on military spending in developing countries, so Anke and I decided to do it ourselves. Because we were new to this question it took a long time to get it right: we finally published the results in 2007. No sooner had we done so than President Arias of Costa Rica, the Nobel Peace laureate, asked us to develop their implications as a support for his initiative for coordinated reductions in military spending. Costa Rica has led the world in virtually eliminating military spending, and we were delighted to be able to add our evidence to support his efforts.

Governments are not exactly forthcoming with data on their military spending. That is scarcely surprising, but it makes the task of analyzing their spending rather harder. I almost persuaded the American government to provide its own private estimates of the spending levels of other countries, but it balked at this. Instead we relied upon the estimates made each year by the Stockholm Peace Research Institute (SIPRI). We decided to measure spending as a proportion of national income: averaged across all the countries of the world for which SIPRI has data, and the forty-year period from 1960 to 1999, the global average was 3.4 percent. Expressed as a percentage it is a small number, but expressed in dollars it is large: by 2006 it had grown to $1.2 trillion. That is around ten times the global aid budget. The countries of the bottom billion alone are spending around $9 billion. This compares with their total aid inflows of only around $34 billion. Our question was why did some countries spend a higher proportion than others, and why was spending higher at some times than at others: the highest proportion we found was 46 percent of income, and the lowest a mere 0.1 percent.

We started with the obvious and gradually got more elaborate. The most blindingly obvious reason for a government spending a lot on the military was if it was fighting a war against some other government. I figured that if we did not find that in the data we should give up and work on something else. There it was in the data: controlling for everything else, if a country is engaged in an international war its military spending increases by 1.5 percentage points of GDP. However, much as you might be forgiven for thinking otherwise, international wars are now so rare that this reason for military spending accounts for only a tiny proportion of global spending: most spending occurs during peacetime.

Just because a country is not currently at war with some other country does not mean that it regards itself as being free of external threats. We scratched our heads trying to think of a good observable proxy for the perception of an external threat. One idea we came up with was once bitten, twice shy. Expressed more professionally, if a country had once had to fight an external war, maybe it would be more fearful that it would have to do so again. Perhaps the country had a particularly dangerous neighbor, maybe it had a particularly aggressive leadership, or perhaps it saw itself as an international policeman, riding to the rescue of distressed regimes. We decided to try it, focusing on the history of warfare since the end of the Second World War. Sure enough, once a country had been engaged in an international war, its military spending was permanently higher by around 1.8 percentage points. We tried to see whether this faded with time. Presumably at some point it does, but we could not find any such tendency: as far as we could see, wars fought years ago were still leaving their legacy in the form of higher spending. If this is right, a disturbing implication is that much of the costs of an international war accrue after it is over: the society continues to be burdened with higher military spending.

A previous war is one reasonable proxy for a perceived external threat, but we decided to try another one that was even more obvious and rather neatly complemented it: the Cold War. The Cold War was evidently a period of perceived threat, but unlike our previous proxy, it was a threat that did not materialize. Further, it had a clear ending, namely the collapse of the Soviet Union. The end of the Cold War therefore constitutes a revealing natural experiment for the coordinated removal of a perceived threat. As you will see, such an experiment is useful: it simulates the effect of ending the cold wars in Lilliput that bedevil the societies of the bottom billion. So what happened once the Cold War ended: did it show up in global levels of military spending? It most certainly did: with the end of the Cold War global military spending fell by an astonishing 35 percent. The collapse of the Soviet Union delivered a huge global peace dividend.

The nature of the threat during the Cold War was, however, unusual: America and the Soviet Union could threaten each other despite the fact that they did not have a common border. This was, of course, due to nuclear missiles. Pretty well all other external threats come from neighbors: no border, no genuine threat. Even the sub-sequent proliferation of nuclear missiles has barely changed this state of affairs: it is because India and Pakistan fear each other as a result of their common border that they now point nuclear missiles at each other.

How big a threat is your neighbor? Well, other things equal, it depends upon how much they are spending on the military. There is a whole body of fancy economic theory that has modeled warfare. Economics has a knack for ugly terminology, and these models are termed contest success functions. The punch line of this work is that if your enemy spends more, then it is wise for you to spend more. You might have sensed that that was quite likely, but you will doubt-less be relieved to know that economists have done the mathematics to prove it. So, armed with the cast-iron certainty that comes from a theorem, we decided to investigate whether there really were arms races in Lilliput. First we had to get countries organized according to who their neighbors were. We found a data set that purported to do this, cleaned it up so that China no longer bordered on Uganda, and got to work. Incidentally, this tells you why our sort of research requires patience: it is necessary to check, check, and check again to guard against the pitfalls you fear. The world had the decency to conform to the theorem: controlling for everything else, if the neighbors spent more, then so did the country itself.

We haven't quite done with proxies for external threats. In fact, there is one more that is utterly obvious. If you were free to chose which neighbor you faced, would you be more frightened of China or Bhutan? Never mind the politics, or the share of military spending in income, China is intrinsically more of a threat because it is so much bigger. More populous countries systematically spend a smaller share of income on the military. This is an instance of a proposition that is going to loom large throughout this book: secu-rity is subject to economies of scale. Big may not be beautiful, but it is safe: small is dangerous, and expensive.

External threats have historically dominated our thinking on defense because of the horrors of international warfare that characterized the last century. There are entire academic departments of international relations that studied the subject. But actually, international warfare is largely a thing of the past. The main drivers of military spending in the countries of the bottom billion are now to be found within their own societies. The threats are internal, not external.

The most obvious internal security threat for which an army might be useful is to counter a rebellion. Sure enough, if a government is engaged in fighting a civil war, its military spending leaps by around 1 percentage point of national income. Civil wars are a lot more common than international wars, and on average they last more than ten times as long. So this form of warfare is more important as a driver of military spending in the bottom billion than is international warfare. Anke and I estimated that for Africa it was about twice as important.

But even civil war is not that common. As is the case with international war, governments spend most of their time fearing it rather than fighting it. Anke and I developed a model of the risk of rebellion and used the estimates to see whether governments responded to a heightened risk of rebellion with increased spending. We found a large effect: governments tried to buy security from the threat of rebellion by building a big army.

This is a case of out of the frying pan into the fire. A big army is not just a source of defense, it is an interest group. The nearest we got to the military-industrial complex was to investigate whether the military was concerned to look after itself. Quite commonly, professions tend to lobby in their own interests. Plato had the brilliant idea that the ideal government would be composed of philosopher kings: government by professors. Unfortunately, Plato's splendid idea has not been implemented with sufficient frequency for it to be amenable to statistical testing, but I hazard that a government of professors would spend significantly more on universities.

Whereas professors have seldom run governments, with generals it is an entirely different story. Sometimes citizens freely choose to elect generals because they are war heroes: American citizens elected General Eisenhower. But more commonly, generals come to power by a somewhat different route: they elect themselves. Military rule happens with sufficient frequency that its consequences for military spending are amenable to statistical testing: does the military do for military expenditure what I expect professors would do for university expenditure? They most certainly do. Following a coup d'état military spending leaps, and quite generally, military regimes spend much more on the military, even controlling for the security risks. Of course, this may well not be ill motivated. Generals, colonels, and what have you may not be driven by squalid ideas such as "It's our turn at the trough." They may be thinking, "At last we can give the defense of our nation the priority it truly warrants." So the phenomenon of the military spending on itself may be well motivated. I have come to doubt the efficacy of dissecting motives: I prefer to look at consequences. Leaders can deliver dire consequences despite fine motivations.

So far I have considered military spending from the perspective of what governments might feel they need to spend, or would like to spend. But as with any other type of expenditure, it has to be tailored to what can be afforded. Although we measured military spending as a share of income, this still leaves out a lot of affordability considerations. One potentially important consideration is how rich the society is. Rich people spend a higher share of their income on luxuries than poor people. This is not a moralizing statement, *it is a definition*. Economists define luxuries as those items of expenditure that increase more than proportionately as income rises. The opposite of luxuries is necessities. Things that are necessities have to be bought even if income is low: that is why food purchases form a much larger share of spending for poor people than for rich people. The current surge in world food prices is an irritant to the budget of a household with a decent income, but devastating for the world's poor: food is half of their expenditure.

So is military spending a necessity or a luxury? Well, we know from politicians that military security is the most vital priority: they tell us often enough. In economic language things that are vital priorities are necessities: you buy them even if it means sacrificing the purchase of things that are less necessary. So military spending must be a necessity. It is one of those propositions that we know must be true, but nevertheless it is best to look: known truths, like theorems, sometimes turn out to be wrong. Indeed, it is wrong. Military spending came out as a clear luxury: it increases much more than proportionately with income. In one sense this is good news: the poorest countries tend to spend a lower proportion of their income on the military. But the unfortunate aspect is that with global growth military spending is going to loom larger and larger. While politicians tell us how necessary it is, they behave as though it were a government luxury good.

There was another aspect of finance that we decided to explore. Low-income countries get substantial amounts of aid. We wondered whether any of this inadvertently financed the military. Any such effect would be inadvertent because aid for development is distinguished very sharply from aid for the military: it is meant to pay for education, infrastructure, and suchlike. Aid intended for military support is recorded separately. Back in the 1960s the United States gave about equal amounts of money for development aid and for military aid. Gradually the balance has shifted: now military aid is far less than development aid. This change in priorities was not foolish: as you will see, security comes with development rather than with guns.

Whether aid leaks into military spending is an easy question to pose but a difficult one to answer. The difficulty is in sorting out causality. We want to test whether aid causes military spending, but it is quite likely that causality also runs in the other direction.

Those governments that choose to have high military spending may get less aid because donors disapprove of their priorities. To test for whether aid leaks into military budgets, we need to focus on variations in aid that are not caused by donor reactions to military spending. Economists realized how to do this in 2003, and we followed the approach that has now become standard. It relies upon the fact that different donors give money to different countries, largely for historical reasons. The Italians give to Ethiopia, a former colony, and the French give to Cote d'Ivoire. Further, national aid budgets go up and down according to the national economic cycle. So, when the Italian economy booms relative to the French, Ethiopia tends to get an aid windfall relative to that of Cote d'Ivoire. Crucially, this is not caused by anything happening in either country, so if military spending goes up in Ethiopia and down in Cote d'Ivoire, it is either a coincidence or aid has changed military spending. Coincidence is always a possibility, but it becomes a small possibility if the number of observations is large: this is what is meant by statistical significance.

So, after all that, what did we find? We found that aid does indeed leak into military spending: on average around 11 percent of aid finds its way into the military budget.

There are many ways in which this might happen. The most evident is that aid that is ostensibly earmarked for some particular expenditure in fact releases the money that the government would otherwise have had to use for that expenditure. The only way to avoid this consequence of aid would be for the donor to insist on items of expenditure that the government categorically does not want. Yet over the past decade there has, quite reasonably, been a massive shift in the philosophy of aid toward country ownership. The government itself sets the agenda for how the aid is earmarked. So it has probably become even easier for governments to use part of the aid flow for military spending. If 11 percent of aid leaks into military spending, since total aid to the countries of the bottom bil-

lion is $34 billion, the inadvertent leakage into military spending
totals $3.7 billion. Since total military spending of the bottom bil-
lion is around $9 billion, this implies that around 40 percent of the
military budgets of the governments of the bottom billion is being
financed by aid. Hopefully, this may overstate the problem. Even if
on average the leakage s 11 percent, if donors manage to skew aid to
the countries where leakages are below the average, the total would
be less than 40 percent.

If we combine this result with the evidence on neighborhood
arms races, it has a potentially disturbing implication. Maybe aid is
inadvertently financing arms races in Lilliput. As we are about to
see, it has a yet more disturbing implication in the context of post-
conflict societies.

Do guns deter civil war? It may well be that once a rebellion
is under way high military spending can squash it, but prior deterrence is potentially a different matter. Whether guns deter is another
of the questions that is easier to pose than to answer. Since the need
for deterrence is one reason for having plenty of guns, it is possible
to get into a statistical muddle. High spending is likely to be associated with a high risk even if in fact it reduces risk. You might recognize this two-way causality as being the same underlying problem
as that posed by aid and military spending. The underlying solution
is the same: find something that influences spending but does not
otherwise affect the risk of war. In the parlance of economics, a subject desperate to look scientific, such influences are termed instruments. In principle, from the differences in military spending that
are only due to such influences, it should be possible to tell whether
the spending *causes* a change in risk: that is, whether it deters.

We followed this approach and concluded that military spending did not deter: in fact it did not seem to have any discernible
effect. This may have been because the instruments were not good
enough: in social science it is far more difficult to show convincingly
that something does not matter than to show that it does matter. But

we went one step further, and this was, to my mind, rather more convincing. As you have seen, the most risky environments are the post-conflict situations. We asked whether the military spending of the government was *differentially* effective in such situations. We found that there was indeed a significant difference, but it was perverse. Far from deterring violence, high military spending by a post-conflict government provoked it.

That is where our work had got to by the time that Anke was due to give birth to Henry: we hastily packaged it up and got it published. I could try to pretend that our research is an entirely ordered and coherent sequence of steps along which we march to discoveries. It should be, but it isn't. Somehow I managed to work on the question of post-conflict risks along two parallel tracks, producing two different journal articles on distinct but related issues. I realized that our analysis of military spending in the post-conflict context had inadvertently omitted other influences that are specific to such situations, most notably peacekeeping. Evidently the two analyses should be merged. With Havard Hegre, a young Norwegian political scientist, I combined them. We again explored the effects of the government's post-conflict military spending, but now alongside peacekeeping. To my relief the result survived: high military spending significantly increased the risk of further conflict.

As to why it has this effect, we have to leap beyond what statistics can tell us and try intelligent speculation. My own guess is that the decision of the government to spend on the military inadvertently signals to citizens that it is planning to turn nasty, and that this signal forewarns those rebels who have recently put down their arms that they were unwise to do so.

I have just taken you through two distinct results: aid leaks into military spending, and in post-conflict situations military spending increases the risk of reversion to war. Now for the dilemma: aid is highly useful in post-conflict economic recovery, which in turn brings down the risk of further conflict. Putting all this together, aid in post-conflict situations is currently a two-edged sword, restoring the economy but inadvertently inflating dangerous spending.

So far I have focused on government military spending, but how do the rebels get their armaments? One route is that hostile governments buy guns for the rebel groups that oppose neighboring governments. But this is not the only way in which government guns find their way into informal usage. Guns purchased for official use leak into the hands of rebels.

The weapon of choice for any self-respecting rebel movement is the Kalashnikov. One reason for this is that it is such a simple weapon that it seldom goes wrong, needs little maintenance, and so can be entrusted to someone who is pretty clueless. This is important because rebel recruits are generally young, ill educated, and amateur. The other reason that Kalashnikovs are popular is that they are cheap. That in turn is because the Soviet Union produced vast quantities of them and licensed their production in some of its satellites. The latest insightful political leader to set up a Kalashnikov factory is President Chavez of Venezuela. Presumably he plans to present them to friendly states.

Economists are particularly interested in prices. I remember one distinguished economist saying to me when I was a young researcher: "It's all we've got." What he meant was that the heartland of economics is its theory of behavior: we assume that people and firms try to maximize something, subject to whatever constraints they face. The key constraint they face is the prices. Armed with this insight, economists can predict how behavior will change when the constraints change. If a price falls, people will buy more of the good. So if Kalashnikovs are cheap, rebel groups will buy more of them. Political scientists do not share this fascination with price, and they are the profession that dominates the study of violent conflict. So the data on guns were all done in terms of quantities: there were plenty of estimates of how many guns were being traded, but no figures on their prices.

For some years I had been trying to find price data: instead I found Philip Killicoat, an Australian graduate student. He volunteered to build a data set on the price of Kalashnikovs on the informal market around the world, country by country and year by year. This took real ingenuity, but he had it. After seven months he had three hundred observations, which was enough for statistical analysis. We were in business. Even a cursory look at the numbers showed interesting patterns: a secondhand Kalashnikov was only about half as expensive in Africa as in the rest of the world. So our research shaped itself around this finding: why were Kalashnikovs so much cheaper in Africa than elsewhere, and did it matter? Phil faced an examination on his work: generally, economics departments are not particularly impressed by the importance of a question; they are concerned about whether the student has applied the latest techniques to the highest possible standards. Somehow, despite being the first Oxford student in half a century to go from zero rowing experience to being selected for the university's team, and spending seven months on data gathering, Phil also managed to gain a distinction grade on his thesis: you can be reasonably confident that his results are defensible.

So why are secondhand Kalashnikovs cheap in Africa? The answer is that the key source is leakage from government armies. Government soldiers are usually very badly paid, and so they are tempted to sell their guns or steal from stockpiles. Government armies buy Kalashnikovs most vigorously when they are fighting a rebellion. So the guns are officially imported into Africa, stolen, and so become illegal, but cannot easily be re-exported to the markets in which secondhand Kalashnikovs fetch a high price. That is because to export the guns out of Africa they have to be imported into countries that generally have sufficiently good border controls to make it difficult. But the guns do not just stay in the country whose government first imported them to Africa. Africa's internal borders are highly porous, and so the cheap guns slosh around the continent going to where there is currently demand: which means wherever there is a war. The next question was "Does it matter?" By this Phil meant whether cheap guns increased the risk of rebellion. According to simple economic theory it ought to do so: cheap guns should make rebellion easier and so more likely, or, stated slightly differently, small rebel groups would tend to buy more guns as a result of their being cheap, and so be more likely to scale up their violence to the point at which it met the definition of a civil war. I have to admit that I thought that this was a pretty demanding question to ask of the data. In the event, Phil found a significant effect: cheap guns increased the risk of civil war. When Phil went back to Australia he generously deposited all his data with the Peace Research Institute, Oslo, where it sits waiting for some other enterprising student to update and expand it.

One implication of Phil's work is that dangerous countries make for dangerous regions. Another is that if Africa's internal borders are porous, the way to curtail armaments reaching the current danger points is to curtail arms flows to the entire region. There are two ways of going about such curtailment, neither easy. One is to address the inadvertent leakage of aid into the purchase of armaments: squeezing the finance for armaments should reduce their inflow. The other is to attempt to impose quantitative restrictions on the trade flows.

Each of these approaches might seem impertinent: other regions buy plenty of armaments, why shouldn't Africa? Before Africans and sympathizers put on the comfortable clothes of indignation, step back for a moment and think of Africa's own interest. Quite manifestly, Africa no longer faces a military threat external to the region: all its threats are internal, either threats between neighbors or fears of rebellion. Threats from neighbors place governments in a prisoner's dilemma. Although each country's increase in its military spending makes it feel more secure, it does so by making its neighbors feel less secure. As I discussed, the neighbors respond to this by increasing their own military spending. Arms races in Lilliput are a menace for the region as a whole. Further, recall that military spending is higher when countries are small. Africa is divided into fifty-four countries, despite having a total population far less than that of India. And so the waste that comes from military spending is more of a problem in Africa than in regions with fewer countries.

The solution to a prisoner's dilemma is cooperation. Africa needs collective action to curtail its arms spending. But the problem with cooperation is enforcement: each government has an individual interest in encouraging its neighbors to coordinate reductions in their military spending while not doing so itself. There are various ways of securing cooperation, but the most straightforward is to persuade a neutral policeman to enforce it. The policeman could be the donors limiting leakage from aid, or it could be the United Nations imposing effective embargoes on arms exports. But in one shape or form Africa should be searching for a policeman. As I explained, the donors have recently reshaped aid allocations so that it is even more difficult to prevent leakages. So how about direct restrictions on arms exports?

IF AID IS INADVERTENTLY FINANCING arms purchases, and if cheap guns make civil war more likely, one possible remedy might be to limit the flow of armaments into dangerous places. Fortunately, the countries most at risk of civil war are not sufficiently industrialized to have an arms industry, so that curtailing the trade might indeed limit the availability of guns. In recent years sufficient popular pressure has built up behind this idea that from time to time it is implemented. For example, during the standoff in Cote d'Ivoire between the government in the south and a rebel group in the north, the United Nations placed an embargo on arms shipments to either side. Are such arms embargoes effective? The work I am going to describe is by Stefano DellaVigna and Eliana La Ferrara.

They reasoned that if anybody knew what was really going on, it was the people who were investing their money in companies that manufactured armaments. Of course, not everyone who buys shares in a company knows much about how it is performing. But it only needs a few people to take the time and energy to find out in order for the share price to be affected: if they discover that the company is being hurt by an embargo, they will sell their shares and the price of the stock will fall. Using this reasoning DellaVigna and La Ferrara got information on which armaments companies were exporting to the country prior to the embargo. They then checked what the announcement of the embargo did to the subsequent price of the stocks.

What they found must at first have seemed confusing: although the stock prices duly fell for some companies, for others prices actually went up. Was this just noise: random movements in stock prices? It turned out that it was not random. The stock prices of armaments companies based in the OECD—that is, the developed countries—fell significantly as a result of an embargo. But the stock prices of armaments companies based outside the OECD rose. They realized that the most likely explanation for this was that these latter companies were breaking the embargo and, in the process, profiting from the absence of competition from the OECD companies.

They concluded from that analysis not that arms embargoes cannot work, but that there is a simple method of policing them more effectively. Suspicions should be raised if an embargoed company's stock price rises as a result of the embargo. It is a simple idea with a lot of potential that demonstrates the payoff to statistical research.

ONE KEY CONCLUSION FROM ALL this is that military spending is likely to be excessive, driven up in an arms race spiral, and so be a regional public bad. Collectively, the countries of the bottom billion are spending around $9 billion on the military, of which up to 40 percent is being financed by donors. Similarly, in regions where borders are porous, a profusion of guns purchased by the government of one country gradually seem to leak onto the informal markets of neighboring countries. These cheap guns on informal markets increase the risk of civil war. The final menace is that in post-conflict societies, which are usually big military spenders, the military is counterproductive, provoking the very risk that it is meant to deter.

Not only is military spending excessive but aid is paying for it. If the international community is minded to put matters right it has two potential strategies: quantitative restrictions on arms purchases or an incentive through linking aid allocations to the chosen level of military spending. Arms embargoes can be made to work despite their past lack of success. Guns are fueling the fire of political violence, and there is a need for them to be curbed.

Chapter 5

WARS: THE POLITICAL ECONOMY OF DESTRUCTION

WHY ARE SOME PLACES PRONE to war? Iraq has deeply confused how people think about twenty-first-century war. The war in Iraq is not a guide to the future; it is a rerun of a phase in world history that is essentially over. Iraq started with an international invasion. So did the two world wars, Napoleon's wars, the Crimean War, the Franco-Prussian War, and the other great set pieces of military history. In the twenty-first century international invasions are going to be infrequent. The wars that will fill our television screens this century will be civil wars, not international wars. Of course, there were civil wars in the nineteenth century as well as invasions, but even the civil wars of the nineteenth century are a hopeless guide to what warfare will be like. The major civil war of the nineteenth century was the American Civil War. This was in form, if not in law, an international war: one alliance of states fought another alliance of states, each with its own recognized territory, government, and army. It's history.

Future civil wars will take the form of a government pitted against a private extralegal military grouping. They will variously be called rebels, terrorists, freedom fighters, or gangsters, but their essential characteristic will be the same. These wars will also be a throwback, but to a very different period of history: the time before nation-states cohered.

To rephrase the question, why are some countries more at risk of civil war than others? If we could answer this question we might be able to do something about it: some of the factors that elevate the risk of civil war might be things that could readily be put right. I think I have the answers to this question, although I am not fully sure of them, even after studying the causes of civil wars for some years. My approach is statistical: I take over all the countries in the world for as long a period as possible and try to find what accounts for why civil wars break out in some places at some times but not in other places at other times: why some places are dangerous.

The core of my approach is to try to predict whether a country has an outbreak of civil war on the basis of its characteristics prior to the war. There are many pitfalls in this approach, but the key problem is the lack of data. Records of the civil wars themselves are not the problem. Astonishingly, a small team at the University of Michigan, the university that pioneered the quantitative analysis of political phenomena, has built a record of all the world's civil wars since 1815. There is even now a rival list built up in Scandinavia. But for most of this period there are too little other data to match against these outbreaks of civil war to try to explain them: for example, for most countries reasonable economic data do not exist prior to 1960. Even if the data were available, prior to 1960, for many years, virtually all the low-income world was in empires that kept the lid on internal conflict. The period prior to empires would probably be revealing, but there are as yet too few hard numbers for my approach to be feasible. Even for the period post-1960 the countries that are most likely to have a civil war also tend to be those least likely to have adequate data on other characteristics: they are the dots and blanks in the global tables produced by the international organizations. Fortunately, time has been on my side.

When Anke and I first tried the approach, in the late 1990s, we could muster only twenty-three civil wars to explain. This was pretty close to being hopeless. By the time of our next effort, published in 2004, we were up to fifty-three civil wars and around 550 episodes during which a civil war could have occurred. This was an improvement but it was still far from ideal. In our most recent work we have been joined by Dominic Rohner, and time has helped in three distinct ways. The most obvious way is that there has been more time in which civil wars have occurred, or more encouragingly, might have occurred but didn't. Anke and I work in five-year episodes, and whereas our previous analysis took the story only up to the end of 1999, we are now able to take it up to the end of 2004. Indeed, this is a very strategic additional five years because it was the period of a major effort by the international community to settle wars, and so we can test whether it also reduced the incidence of outbreaks. But time has helped us by more than just this. Scholars have been hyperactive in quantifying phenomena that were not previously measured, and at filling in the gaps in previous estimates, so that our data for the past are now much nearer to being complete.

The third way in which time has helped is the embarrassing one: we have got better at doing the work. We realized that we could use a fancy statistical program that fills in the blanks of missing data by randomly assigning a range of different numbers. I had always been resistant to using make-believe numbers, but the advantage of this approach was that it filled in each missing number with several different possibilities, one at a time. Using these numbers in turn, you could then see how robust the results were to the possibility that the missing number would have taken these values. We used this to check on our core results derived only from those numbers that were genuine.

Another way in which we got smarter was better to control for reverse causality or common causality, the current obsession in professional economics. Take one of our core results, that low-income countries are more likely to have outbreaks of civil war. Is this more than just an association between low income and civil war: the two phenomena tending to occur together but the first not causing the second? The step from association to causality is tricky, and I will try to spell out why. A rudimentary way of getting from association to causality is the sequence in which events happen. If the low income occurs before the outbreak of war, then this suggests that causality runs from income to war. But is this enough? There are three ways in which this step from association to causality could be mistaken.

One is that the civil war could be anticipated. If you know that you are living in a country at risk of violent conflict, then you are less likely to invest. And so the country will be tend to be poor due to the war even prior to the outbreak of the war. But low income has not caused the war; instead the prospect of the war has caused low income. Another is that the country might have some characteristic not included in our analysis that keeps causing civil wars: for example, Jonas Savimbi launched two civil wars in Angola. Since civil war destroys the economy, by the time of the second civil war, Angola was poor: low income preceded that second civil war even if it did not cause it. So something offstage, namely Savimbi, kept causing wars, and the first war reduced income. The final one is that some phenomena are likely both to lower income and to increase the risk of civil war. Bad governance might destroy the economy and give people cause to rebel. So just because low income occurs before war is not enough to conclude that it causes war.

Gradually, economists have become better at guarding against these problems, introducing steps that leave fewer and fewer ambiguities. In our new work we have used more of these safeguards: indeed, having more observations makes it easier to do so because the safeguards generally need large samples. To give you a taste of the safeguards, we got rid of the Savimbi problem by restricting the analysis to the prediction of first-time civil wars. In part, we reduced the bad governance problem by controlling for it, and by including as many characteristics as possible in the analysis. We addressed the problem of anticipated conflict by replacing the actual level of income with the level predicted by a few characteristics that influence income but do not otherwise affect proneness to civil war: in principle this approach should also fix the previous problem. Even with these safeguards there is room for doubt, but at least we now have results based on comprehensive data—at its maximum more than sixteen hundred episodes during which eighty-four civil wars broke out. This may not be as good as it can get, there is always room for improvement, but the results are worth serious attention.

Although we are economists we have tried to be agnostic as to what might explain proneness to civil war, and so we have included a wide range of possible causes drawn from across the social sciences. In addition to various characteristics of the economy, these include aspects of the country's history, its geography, its social composition, and its polity. Let me be clear about what we do not include: we are not interested in the personalities and immediate political circumstances leading to the conflict. All wars have multiple causes: one reason that Saddam Hussein invaded Kuwait was that the Kuwaiti leadership was sufficiently impolite to doubt whether he had been born in wedlock. Such things matter for a proper understanding of any particular war but clutter up and detract from our understanding of civil war as a phenomenon. In trying to prevent war I suppose that it is useful to know that insulting psychopaths is not a good idea, but my approach has been to try to find structural characteristics that expose a country to risks and could, over time, be changed. So let's get started: what actually caused these eighty-four civil wars?

The economy matters. Low-income countries are significantly more at risk even when we control for as many of the possible spurious interpretations as possible: poor is dangerous. Nor is it just the level of income: it is also the rate of growth. Given the level of income, societies that are growing faster per capita are significantly less at risk of violent conflict than societies that are stagnant or in decline. In one sense this is hopeful: it tells us that economic development is peace-promoting. I have no patience with the romantics who believe that we can build peaceful societies by arresting economic growth: the vision of the restoration of Eden. I think that the truth is quite the opposite.

The statistics of the world post-1960 are supported by the deep historical evidence of the societies of early history. These impoverished societies were extremely violent, as Azar Gat has now brilliantly shown in *War in Human Civilization*. Economic development is a key remedy to violence. The truly difficult issue about the peace-enhancing effects of economic development is to sort out which of a number of possible routes might account for it. I suspect that there is no single magic route that could be isolated and promoted distinct from overall economic development. My guess is that there are multiple routes, such as jobs, education, hope, a sense of having something to lose, and more effective state security services, all of which contribute something.

The level and growth of income are not the only aspects of the economy that matter for violence. Dependence on natural resources also increases risks. This proposition is supported by the grim evidence of resource wars: timber in Liberia, diamonds in Sierra Leone, a wonderland of minerals in the Congo. It is also now supported by statistical analysis of where violence occurs within countries. For example, in Angola the violence tended to be concentrated in the diamond areas. It is also evident why natural resources might increase proneness to violence. They provide a ready source of finance for rebel groups, they provide a honey pot to fight over, and they enable the government to function without taxing the incomes of citizens, which gradually detaches it from what citizens want.

Nevertheless, this is probably the most controversial of our results: some scholars have argued that it is purely an oil effect, and others that we have run afoul of a particular variant of reverse causality. Anke and I have learned not to be proud: over the years I have been wrong sufficiently often to have had the presumption of infallibility knocked out of me. With our new data we duly tested whether oil was the real story: as far as we can see it isn't. We do, however, find that with sufficient natural resources a country becomes safe. Saudi Arabia and the other superrich Gulf States are peaceful: they can afford good security systems and they can afford to buy off all potential opponents. Indeed, just this ambiguity—some resources increasing risks but sufficient abundance reducing them—is predicted by the sophisticated recent theoretical work of Francesco Caselli of the London School of Economics.

The reverse causality problem is trickier. It arises because we measure resource dependence by the ratio of primary commodities to income. That inevitably creates a problem because countries that for whatever reason have low income will tend to have a high share of primary commodity exports, simply because income, the denominator, is small. Some scholars have recently tried to get around this problem by replacing our measure with a newly available measure: the value of natural resource reserves. The World Bank released estimates, country by country, giving a snapshot for the year 2000. Unfortunately this runs into another form of the reverse causality problem. Any estimate of natural resource reserves depends upon what resource extraction companies have found through prospecting. Prospecting is costly, and so the value of proven reserves is an economic concept as much as a geological one. It is only worth doing in places where the company's rights of extraction are secure. Between 1960 and 2000, prospecting thus tended to avoid societies that were at civil war, and also those places where there was a serious risk of war. Think what this implies. The places with few proven natural resource reserves in 2000 will tend to be those with the worst prior history of civil war. The scholars who followed this approach duly announced with a confident fanfare that possessing large endowments of natural resources actually makes a society safer. The problem of reverse causality has recently been overcome by Tim Besley and Thorsten Persson. They investigate whether increases in commodity prices affect the risk of civil war in commodity exporters. Consistent with our own results, they find that the risks are increased. But they find an important qualification: the risks are not increased if the quality of democracy is sufficiently high. As with elections and reform, democracy is a force for good as long as it is more than a façade.

So much for the economy; let's turn to history. The aspect of a country's history that most commonly excites interest when it comes to explaining a civil war is its colonial experience. Understandably, many people in developed countries find it convenient to emphasize the guilt of their own societies, and equally, many people in developing countries want to avoid any impression that the violence of their societies is a consequence of characteristics within those societies. So there is a ready demand for evidence that colonialism is responsible for the subsequent violence. Unfortunately, Anke and I cannot find evidence that supports this contention. Neither the length of time that has elapsed since independence, nor the particular former colonial power, seems to matter. I do not want to push this too far: it is quite evident that Portuguese decolonization was disastrous. Angola, Mozambique, East Timor all went straight into civil war. But the Portuguese empire was relatively small, and neither the British nor the French empire, which were the two major ones, shows any distinctive patterns. The empire-free countries of Ethiopia and Liberia both eventually collapsed into terrible civil wars. I want to stress that this is not to exonerate colonialism: I am not an apologist. But blaming colonialism for civil war is a costly illusion because it detracts from the focus on its real causes, which are often things that can still be changed. It may make many people feel better, but it inhibits action.

The other aspect of history that many scholars have got excited about is the Cold War. Quite evidently, in some instances civil wars were aided and abetted by each side. As Niall Ferguson has pithily expressed it, what people expected would be the Third World War turned out to be third-world wars. But even the effect of the Cold War is controversial. While it is clear that the superpowers intervened in civil wars, it is less clear that they caused them. Indeed, they may even have had an offsetting effect: if any petty war had the potential to scale up into the Third World War, maybe the superpowers tried to prevent conflicts occurring. We tested for this by investigating whether the post–Cold War period has had a significantly different incidence of the outbreak of civil war than we might otherwise expect. Basically, it doesn't. There was a brief, significant surge in new outbreaks of violence in the first few years after the end of the Cold War, but from 1995 onward the world has been back to normal. The third world's wars were not, in general, caused by fears of the Third World War.

The one aspect of history that really seems to matter is a previous history of civil war. Once a country has had a civil war, it is much more likely to have another war. However, this is ripe territory for the problem of common causality. Suppose that there is some characteristic of the country that makes it prone to violence but that we have missed: perhaps the people are just inherently violent. Statistically this will appear as one war causing another, whereas actually the same underlying factor is causing all of them. We got around this problem by measuring the number of years since the last civil war and testing whether that, or the mere fact of having had a previous civil war, was decisive. It turned out that it was only the length of time since the previous war that mattered: the risk of further conflict gradually declined with the passage of time. This looked more like a risk of violence caused by the gradually decaying effects of previous violence than by something underlying and constant.

So much for history: how about the structure of the society? Perhaps the most important aspect of social structure that we investigated was the effects of ethnic and religious divisions. This is the point at which I get mud on my face because our previous results got overturned by the new data. We had previously found that ethnic and religious diversity had two opposing effects. We now find that the relationship is more straightforward: diversity increases the risk of violence. As far as we can tell, ethnic and religious diversity compound each other.

Another aspect of the social structure that seems to affect the risk of violence is the proportion of young men in the population: young men, defined as those aged between fifteen and twenty-nine, are dangerous. I suppose this is not surprising: violent rebellions are seldom staffed by elderly ladies. The effect is large: young men appear to be very dangerous. A doubling in the proportion increases the risk of conflict from around 5 percent over a five-year period to around 20 percent. However, there are a couple of caveats here. It is very hard to distinguish statistically between societies with many young men and those with many young women: other than in China the two tend to go together. In most rebellions the fighting is done almost exclusively by young men, but not always: famously the Eritrean People's Liberation Front was one-third female. And it is also hard to distinguish between societies with a high proportion of young men of fighting age and those that simply have rapid population growth.

A final aspect of the society that matters is its size. The risk of conflict increases with population, but the relationship is much less than proportionate. A country with a population double that of an otherwise identical country has a risk that a civil war will break out that is only a little higher than its smaller counterpart: specifically it is one-fifth higher. Think about that for a moment: it implies that if two identical countries were merged, then, abstracting from all the nationalism that would of course be provoked, the risk of a civil war breaking out somewhere in the combined territory would fall. Let's say that previously there was a 10 percent risk in each country, so that the risk that there would be a war *in one or the other* of the countries was around 20 percent. Now, the risk of a conflict in the new super-country is only a fifth higher than that in either of the former countries individually: that is, the risk is 12 percent. So the risk of war has *fallen* from 20 percent to 12 percent.

I think that this is because there are scale economies in security, and I think that it matters. Most of the countries that emerged with the dissolution of empires were too small to reap adequate economies of scale in security. This creates a potential tension between the scale economies that could be reaped by mergers between countries and the greater ethnic diversity that would probably be a consequence. At present all the political pressure is for nations to get smaller. Eritrea exited from Ethiopia. East Timor split from Indonesia. Yugoslavia split up into six different countries. Southern Sudan will hold a referendum to decide whether to withdraw from Sudan. Stepping back from the historical particularities of these struggles for nationhood, is the drift in the right direction?

So much for social structure: how about geography? We had already tried to investigate whether particular types of geography were well suited for rebellion. The most promising idea seemed to be that of the safe haven, and two aspects of the landscape seemed likely to facilitate: forests and mountains. Forests were relatively easy to measure: the Food and Agricultural Organization had a measure, country by country. We investigated it and could find no effect. But there was no equivalent measure for mountains. There were crude proxies such as the highest point in the country, but these seemed to miss the point as to what rebel groups would actually find useful: they did not want to perch on the top of Everest; they wanted rugged terrain where government forces would not be able to find them. We tracked down the world's leading geographer on mountainous terrain and commissioned him to build a quantitative measure of the proportion of a country's terrain that could reasonably be judged mountainous. This measure has since become widely used, and in our new work we indeed find it to matter: mountains are dangerous.

Finally, let's get to the politics. Surely that must be where the real causes of violent conflict are to be found. We investigated a range of political science variables, but focused on the one most widely used by political scientists, Polity IV. I have already described the results: in low-income societies democracy is dangerous, and in high-income societies dictatorship is dangerous. Other than this we could find no effects. Many people quite reasonably assume that violent internal conflict is the consequence of political repression, but we simply do not find evidence for this in the data. This does not, of course, mean that repression is all right. Repression is unjust by definition because it denies political rights. But this can be undesirable without making a society more dangerous. And it is danger that is my subject.

It is time to try to make sense of this evidence. This involves an interpretive leap from the statistics. Some interpretations become implausible in the face of the statistical evidence, but more than one interpretation is possible, and my own may be wrong. With that caveat I propose the *feasibility hypothesis*. The feasibility hypothesis is that the key to understanding civil war is to focus on how rebellion happens rather than on what motivates it.

Why focus on the rebels? Does that reveal a pro-government bias? The focus on the rebels is simply because it is the act of rebellion that defines the outbreak of civil war. All governments with the exception of Costa Rica and Iceland have armies, so these cannot be the defining feature. Sometimes a government army attacks its own defenseless citizens, but, disgusting as this is, it is a pogrom, not a civil war. The defining feature of the outbreak of civil war is that the usual monopoly of force held by the government army is challenged: a private organization within the society builds its own army. No government can tolerate the existence of a private army on its soil, and so even if it is the government that fires the first shot, it is the creation of the rebel army that defines the war.

Because of the emphasis upon why the civil war is being fought, it has become natural to focus on what *motivates* the rebel group to form an army. My own past work fed this perspective: a paper entitled "Beyond Greed and Grievance" questioned the conventional view that rebels were motivated by a sense of grievance, introducing the idea that they might also be motivated by greed. But it was essentially a refinement within the motive-based explanation for rebellion. I have now moved on from this view. It seems to me that the key insight into rebellion comes not from asking *why* it happens but *how* it happens. Usually rebellion, at least on a scale needed for civil war, is simply not feasible. The definition of civil war that I have used, which is conventional, is that at least one thousand people are killed in combat per year. On this definition the average civil war lasts around seven years. So we are looking for rebel organizations that can kill and be killed on a large scale and yet survive for years. Rebellion on this scale faces two major hurdles. One is money: a rebellion is going to be expensive. Someone has to pay for the guns, and someone has to pay for the troops.

Often people think that a rebellion is just another form of political protest: people fight when they can't vote. What brought it home to me that rebellion is not simply a variant on other forms of political opposition was a comparison I made between the finances of a medium-size rebel group and a major political party. The rebel group I chose was the Tamil Tigers. As rebel groups go it is not out of the ordinary: northeastern Sri Lanka, where it operates, lacks high-value natural resources; this war is not financed by diamonds. I chose the Tamil Tigers only because, unusually, its finances have been reasonably well studied. Its annual revenue is around $350 million. This is around 28 percent of the GDP of North East Sri Lanka, although most of the money is generated outside Sri Lanka from donations by Tamils abroad.

For a political opposition party I decided to look for a rich one. I chose the British Conservative Party, one of the longest-surviving and most successful political parties in history, which, being on the political right, is able to tap readily into financial support. I chose the election year of 2005, when presumably its revenues were relatively high. This information was more accessible than that about the Tamil Tigers: its revenue was around $50 million. So one of the best-financed political opposition parties in the world had an income one-seventh that of a medium-size rebel movement. Recall that the revenue of the Tigers was 28 percent of the GDP of the area they sought to control; expressed in that way, the British Conservative Party was not one-seventh the size of the Tigers, it was one ten-thousandth. There is no simple passage from political opposition to private army: there is a cliff face in the form of a financial barrier. Most would-be rebels just cannot muster the money regardless of their motivation.

The other hurdle is military. Under most circumstances if a small group of young men arm themselves and oppose the government army, either they confine themselves to the irritant of terrorism aimed primarily against civilians or they die. Only if they are faced by a militarily weak government do they stand much chance of survival. While a rebel leader in Zaire, Laurent-Désiré Kabila, was able to hang on for many years, safe because President Mobutu had undermined all the organs of government, including the army.

So what is the feasibility hypothesis? It is that in explaining whether a rebellion occurs, motivating factors are of little importance compared to the circumstances that determine whether it is feasible. The tough version of the hypothesis, which I am reluctant to adopt but which I suspect is close to the truth, is that where a rebellion is feasible it will occur: the rebel niche will be occupied by some social entrepreneur, although the motivation might be anything across a wide range. Civil war is predominantly studied in political science departments and so naturally enough they interpret the motivation as political: sometimes it surely is, although even political motivations might stray quite some distance from social justice. Even rebellions that look entirely justified can sometimes be called into doubt.

Take the rebellion, or rather rebellions, in Darfur. Quite evidently the government of Sudan is awful, and its conduct during the conflict has been murderous. But at least part of the impetus for the Darfur rebellion was the settlement of the rebellion in the South. The Sudanese People's Liberation Army, which fought the rebellion in the South, won some remarkable concessions from the government in the North: it was allowed to run its own government, it received a substantial share of the oil revenue, gilded by the promise of huge aid inflows from donors, and all this was capped with the promise of a referendum on full independence six years after the onset of peace. No sooner was this deal signed than the contingent from Darfur that had been fighting for the SPLA returned home and launched its own rebellion. You can certainly see why, with that precedent, rebellion might be attractive, at least for its leadership. The top dog would become a president, and the others would become ministers: secession has its rewards. The rebellion is, of course, justified in terms of the atrocious sufferings of the Darfur people. But to date the consequences of the rebellion for the people of Darfur have been catastrophic: surely far worse than any plausible alternative scenario. Either the rebel leadership radically misjudged the consequences of its actions, or it was not genuinely motivated by the welfare of the people of Darfur. When the government was recently coaxed to the negotiating table, the key rebel organizations refused to attend. It is hard to see how a refusal to negotiate can be in the best interests of the people of Darfur.

Sometimes the motivation for rebellion seems to be religious, with the rebel group more akin to the fringe religious groups such as those in Waco or Jonestown, but with the violence turned outward. For many recruits the motivation may well be the lure of violence: only a small minority of any society are psychopathic, but these people are likely to be in the front of the queue for rebellion. Sometimes it might even be sexual. Joseph Kony, the leader of the Lord's Resistance Army in Uganda, has reputedly accumulated sixty wives: perhaps a young man's dream come true?

The statistical results do not prove the feasibility hypothesis, but they are consistent with it. I used the results to simulate the risks of conflict in two hypothetical territories, in one of which rebellion was easier than the other. I varied only five characteristics that seemed to be most readily interpretable as differences in the feasibility of rebellion. One territory was very mountainous, the other was flat: mountains provide safe havens for rebels. One had a high proportion of young men; the other had a low proportion: young men are the recruits on which rebel organizations depend. Both territories had a population of fifty million, but one consisted of a single country whereas the other was split into five identical countries, each of ten million: the small countries would be unable to reap economies of scale in security. One was dependent upon natural resource exports, the other was not: such exports can provide finance for rebellion. One was in Francophone Africa and so benefited from the French security umbrella, the other was not. All the other characteristics were the same and set at the average for all the countries in the analysis. I then predicted the risks for these two countries. The easy-rebellion territory faced a risk of 99 percent that conflict would break out in one or other of its countries during a five-year period: this territory was basically so dangerous that it was condemned to perpetual conflict. The difficult-rebellion territory faced a risk of less than 1 percent: basically it was safe, even over a century, it was highly unlikely to fall into violence.

Dramatic as these differences are, they are not decisive evidence. Most of the differences in characteristics that I have used to construct easy-rebellion and difficult-rebellion countries could instead be interpreted in terms of motivation. For example, I have interpreted the increased risk of rebellion that mountains induce as being because they are safe havens for rebels. But here is an alternative, motive-based explanation. The people living in the mountainous areas of a country are usually poorer than those in other parts of the country. They may storm down from the mountains to redress this grievance: mountains matter, but because they create pockets of grievance. While I do not want to discount such alternative interpretations, I think it is striking that the most obviously grievance-related characteristics such as the polity do not seem to make much difference to risks, whereas these characteristics that at least have plausible interpretations in terms of feasibility have such a large effect.

WHAT DOES CIVIL WAR ACHIEVE? Most obviously, war kills and injures people. Most of the dying is not as a result of battle, but due to sickness. Mass flight takes people into unfamiliar places where they lack natural immunity, and public health systems collapse. Because disease is highly persistent, much of the dying occurs after the war is over.

Also pretty obviously, war is bad for the economy: not only does it destroy the economy of the country itself, it damages the neighboring economies. Again, these effects are highly persistent so that many of the economic costs occur after the war has ended. I estimate that for the typical civil war in a society of the bottom billion, these economic costs alone are the equivalent to losing around two years of income, or some $20 billion. However, I have come to realize that these estimates, though large, grossly understate the true cost.

They make no allowance for the fact that the people affected by violent internal conflict tend disproportionately to be among the poorest and most disadvantaged people in the world. A dollar lost by someone who is poor should be valued more highly than a dollar lost by someone who is better off. The income differential between the typical citizen in the countries of the bottom billion and a typical citizen of the other developing countries is already around one to five. Even within the bottom billion there is a wide range of incomes, with those countries that have recently been in conflict grouped right at the bottom. Not only are the war-prone already the poorest, they are likely to stay the poorest. Since slow growth is itself a significant risk factor in violent conflict, the most violence-prone countries are systematically among the slowest growing.

My estimates of cost also make no allowance for the fact that peace is fundamental to development, so that its absence frustrates all other potential interventions. The vaccination of children or the reliable provision of anti-retroviral drugs is virtually impossible in wartime conditions. This creates weakest-link problems in the provision of global public goods. For example, smallpox was eliminated globally in a country-by-country campaign that was evidently a race against time: until it was eliminated everywhere there was a risk that it would break back out as a global disease. The last country on earth where it was eliminated was Somalia during the 1970s. It would now be impossible to eliminate smallpox: since 1993 Somalia has been a no-go area. The maintenance of peace is thus a logically prior investment that opens the possibility of all other interventions. It is even possible to dress this up in the language and formulas of technical economics. Financial economists now calculate option values. The true return on a liquid asset such as a bank deposit is greater than the interest earned because it enables other investment opportunities to be seized as they arise. Peace also has an option value.

Finally, I have made no allowance for three global spillover effects: crime, disease, and terrorism. Large-scale political violence and the resulting breakdown of the state create territories that have a comparative advantage in international criminality. They provide safe havens both for criminals themselves and for their material activities, such as the storage of illegal commodities, notably drugs. Some 95 percent of hard drug production is concentrated in civil war or post-conflict environments. Civil wars also create the conditions for the spread of disease: the breakdown in public health systems and the mass movement of refugees. Some of this spread of disease affects neighbors, and potentially it can also affect the entire world. One of the explanations for the origin of AIDS for which there is some evidence is that it originated during a civil war. Finally, civil wars appear to assist terrorism. Al Qaida based its training camps in Afghanistan because the absence of a recognized government was convenient. Similarly, the American government finally decided that leaving Somalia without a recognized government was too dangerous, once evidence built up that Al Qaida was relocating there.

Where this leaves us is that the cost of this form of political violence is enormous. Even were it to lead to healthy political change, we would need to ask whether the eventual benefits were worth these massive costs. But the final tragedy of civil war is that it does not usually lead to any such political legacy. If we take as our measure of the polity the Polity IV index, civil war leads not to improvement but to deterioration. Instead, as we have seen, the most likely legacy of a civil war is a further civil war.

IF THE FEASIBILITY HYPOTHESIS IS right it has a powerful implication: violent conflict cannot be prevented by addressing the problems that are likely to motivate it; it can only be prevented by making it more difficult. Whether rebellion is easy or difficult basically comes down to whether rebels have access to guns and money, and whether the state is effective in opposing them. Most of the guns and money that finance rebellion come from outside the societies that are plagued by civil war. The effectiveness of a state increases with its level of development. This gives the international community some scope to reduce the incidence of war. It can squeeze rebel organizations by curtailing the guns and money that presently reach them so readily, and it can try to break the impasse that has frustrated development.

Should the international community try to discourage rebellion? When that iconic poster of Che Guevara first came out, I was a student. For my generation, support for armed struggle in developing countries was a natural extension of our support for liberation movements. But liberation from colonialism and rebellion are not the same thing: one unites the society against an external oppressor; the other divides it against itself. Painful as it is to revise cherished old beliefs, armed struggle is usually development in reverse.

Chapter 6

COUPS: THE UNGUIDED
MISSILE

T HE COUP D'ÉTAT AS A technology of political violence
will play a central role in this book. The ghoulish glam-
our widely associated with political violence has focused
almost exclusively upon rebellion—armed struggle, as it is called by
its aficionados. Rebellions should be turned into history as rapidly
as possible because the consequences of civil war are so dire. But
coups are a different matter. The challenge posed by coups is not to
eliminate them but to harness them. Coups have the ready potential
to deliver what armed struggle was supposed to achieve but seldom
did. But I am not about to give you a eulogy for coups: to date, they
have usually been awful. It is time to look.

Suppose that you are the president of a country that is one of
the bottom billion. Although life for ordinary citizens is tough, fortunately a grateful nation has made your own life remarkably comfortable. In the developed countries presidents have to wait until
they step down and write a best-selling memoir before they come
into money. Even in the more successful developing countries political power often does not lead to wealth: former president Muhatir
of Malaysia—one of the most successfully transformed countries
on the planet—is not a rich man after many years in power. When

President Quett Masire of Botswana, Africa's most successful economy, stepped down, he feared that he might be declared bankrupt. But in the societies of the bottom billion, political leaders have a long tradition of accumulating wealth while in office. Far from offering the prospects of book tours, losing power looks decidedly scary. So evident is the fear of losing office that a public-spirited African businessman, Mo Ibrahim, has now introduced a $5 million prize for African presidents who voluntarily step down. Quite possibly, over time this will indeed change behavior.

Recall that most presidents have learned how to live with the menace of elections. The prospect of facing the electorate once every few years is not what brings presidents out in a cold sweat in the early hours of the morning. What brings on the fear is, ironically, the system that is supposed to keep the country secure: presidents fear a coup d'état from their own army. Since independence, successful African coups have been running at a rate of around two a year. Unlike an election, it could happen at any time of any day or night. If a coup succeeds, sometimes the president can get out in time, but not necessarily. The successful coup leaders who toppled President Doe in Liberia not only tortured him to death but made a video of it. So presidents are right to be scared. This is the analysis in which some of them will be most interested. It is based on work with Benedikt Goderis and Anke Hoeffler.

While the prospect of a coup obviously matters to presidents, it is not clear that it should matter to the rest of us. If the only regimes that are threatened are themselves dictatorships, a coup is not something to feel outraged about: perhaps it is the only way of disciplining dictators. Of course, it would be a different matter if democracies were also threatened by coups. Even if coups replace dictators, as an economist my instinct is to say, "What does it cost?" It is obvious that a civil war is immensely costly, both in the narrow terms of income forgone and in the deeper terms of mortality and the breakdown of social cohesion. But that is because civil wars are prolonged, destructive, and usually indecisive. A coup is a surgical strike: perhaps it is a cheap and effective way of ousting bad regimes. Rather than speculate, I decided to investigate. Since at this point presidents will be tempted to throw the book down in disgust, let me reassure them that soon enough I will get to something of more evident importance: what makes a coup less likely. Impatient presidents can skip to the next section.

To get a sense of the cost of a coup, a useful starting point is its impact on the growth of the economy. We found a clear and straightforward effect of a coup: in the year of the shock it reduced income by around 3.5 percent, but after a couple of years the economy reverted to normal. So, taking the aftereffects into account, this particular cost was around 7 percent of a year's income. We realized that this cost might be merely the tip of an iceberg. Economists have found that political instability is bad for the economy, and coups seemed likely to be an important form of instability. The main costs of coups might not be the brief aftereffects of successful coups, but the consequences of the continuous fear of them. In countries at high risk of a coup d'état, investors may keep away. To investigate whether the risk of a coup is damaging, you first need to estimate the risk. In the process you discover what makes a country prone to a coup, which I will describe shortly. We introduced this risk into our analysis of growth, trying to see whether it reduced economic activity over and above the actual coups themselves. We could not find an effect, and while that does not mean it isn't there, it probably means that any effect is fairly modest.

A price of 7 percent of a year's income is not a cheap way of replacing a government, but if the government is truly terrible and it is replaced by a better one, then it is probably a bargain. Would the Iraqis have paid 7 percent of a year's income to oust Saddam Hussein, thereby avoiding a war? Would the Zimbabweans have paid it to oust President Mugabe, thereby avoiding an economic meltdown and mass emigration? This is an important difference between a

coup and a rebellion. A rebellion with its consequent civil war inflicts such high costs on the society that in my view there should be
a strong presumption that rebellion is undesirable. Armed struggle
may be romantic, but it is usually a menace. It is sometimes argued
that if governments and rebels are equally bad, as I think is often
the case, then the international community should stay neutral. I
strongly disagree with this view. Unless the rebels are unquestion-
ably a whole lot better than the government, then the cost inflicted
on the society for the one-in-five chance that the rebellion will lead
to the government being overthrown is far too high, and so the rebellion should be discouraged. Neutrality is inappropriate when the
issue is war or peace. But coups are a different matter: they have to
be judged predominantly by whether they improve governance.

It is easy to come up with reasons that a coup might improve
matters, and sometimes they clearly do so. The mere threat of a
coup may act as a restraint upon the government. For example, one
of the few instances in which an election in one of the societies of the
bottom billion resulted in the defeat of the incumbent and a change
of president was in Senegal in 2000. The incumbent was persuaded
to accept defeat because the army told him that if he did not, they
would mount a coup. In turn, the Senegalese military had been emboldened by the successful coup d'état in Cote d'Ivoire a few months
earlier. So, in that instance, one country's coup was another country's safeguard of the democratic process.

Not only might the threat of a coup discipline a government,
but *in extremis*, a coup might be the only way of replacing a dysfunctional leader. Colonel Ely Ould Mohamed Vall led a surgical coup
in Mauritania in 2005, promising clean elections in which he would
not be a candidate, and duly kept his promise. The elections, properly conducted, ushered in what currently looks to be an excellent
government. But unfortunately, even good coups that replace terrible rulers can end up further degrading the polity. Emperor Hailie
Selassie of Ethiopia built a regime in which power was entirely con-

centrated in his own person. By 1974 he was a senile octogenarian ruling with disastrous incompetence over the poorest country in the world. Visiting him, the emperor's retired adviser John Spencer was so shocked that he predicted a coup within six weeks. In fact it happened the next day. So far, so good for everyone except the emperor: the coup replaced a senile emperor with a respected general, Aman Andom. But this was not the end of the story. Easy as it would have been to improve on the performance of a senile emperor, the coup ended up producing an even greater catastrophe: Colonel Mengistu Haile Mariam. The general who led the first coup was replaced in a further coup, and the new leader marched the country into a disastrous war, in the process creating one of the world's most repressive and economically ruinous regimes.

Even worse, coups might not be provoked by bad governance but by the opportunistic greed of the army. No sooner had democratic São Tomé discovered oil than the army attempted a coup. The nighttime coup that ousted President Sir Dawda Jawara in the Gambia originated when a group of drunken soldiers decided to go to the Presidency building to demand higher pay and found it undefended. The Thai coup of 2006 deposed a democratically elected government that duly got reelected once citizens were given the chance to vote. So, for the moment, we will have to park the question of whether coups are surgical strikes against bad governments that are cheap at the price, or a menace posed by greedy gunslingers: either might be the norm. I want to get back to what a worried president might want to read.

PRESIDENTS, RESUME READING HERE: I am going to investigate what determines a coup. As usual, my approach is to gather as much data as possible on coups and then try to explain their occurrence statistically. There is a standard international data source on successful coups d'états around the world. This was promising, but then I hit upon a new data set that had been put together by Patrick McGowan, a political scientist in Arizona. His innovation was to have recorded not only the successful coups but the failed attempts, and even those that had never got as far as an attempt but had been nipped in the bud at the stage of a plot. His data were only for Africa, but this still added up to a large number of failed plots, failed attempts, and successful coups. We reasoned that all attempts, whether successful or not, had at some stage been plotted. This gave us an amazing 336 coup plots, of which 191 made it through to the stage of an attempt, and 82 made it all the way to a successful coup. Our task now was to explain what determined each of these stages: why plot; what got a plot through to an attempt; and what made an attempt succeed?

Since the true purpose of this section is not to help worried presidents retain power, but to discuss how to curtail this form of political violence, I will start with the key issue. Does democracy makes coups less likely? Controlling for other influences, unfortunately it does not: coups are at least as likely to break out in democracies as in autocracies. I say "at least" because severe repression significantly reduces the risk of a coup. So precisely when a coup would be most justified, it is least likely to occur. We checked to see why repression made governments safer and found that it enhanced the ability of governments to detect plots. Repressive regimes did not face more plots than other sorts of regimes, but the plots were more likely to be aborted before they could reach the stage of an attempt. Behind that bland statement is the grim reality of repression: torture, fear, agents provocateurs, and spies. They work, which is presumably why they are used so enthusiastically by dictators around the world. It is back to Herodotus and the preemptive weeding of possible opponents. The worried president sets down *Wars, Guns, and Votes* for a moment and jots down a memo on the pad beside his bed: increase the budget for military intelligence.

A second rather disturbing feature of coups is that one leads to another, just as it did in Ethiopia. The baseline risk for a coup attempt in Africa is 4 percent per year. Following a coup attempt the risk of further attempts is greatly elevated: a year after an attempted coup there is a 10 percent risk of a further coup. Evidently, the same arguments that General Andom used to justify his seizure of government by force could be used by Mengistu to justify seizing the government from Andom. The very act of usurping power destroys the defense of legitimacy. Perhaps more potent than the lack of legitimacy is that a coup sets an example. General Andom inadvertently demonstrated to younger army officers that they could transform their lives from the squalid setting of a barracks to the luxury of the presidential palace by one audacious act. Although General Andom was probably motivated by the best interests of his country, Mengistu was probably motivated by interests rather less lofty: soon enough he was being driven around Addis Ababa in a red Cadillac. What is more, if the coup removes the people currently at the top, everyone else has a chance of moving up, so it is relatively easy for new coup leaders to gain support among their colleagues.

Taken together with the previous result, coups are less likely to throw out truly bad governments than to throw out better ones. They are also likely to lead to further coups, each of which incurs costs. Coups are beginning to look less attractive than we might have hoped.

Do ethnic divisions matter? This is one of the relatively few respects in which Africa is distinctive. Usually I find that Africa conforms closely to global patterns of behavior: the outcomes are distinctive only because the characteristics that globally drive behavior are distinctive in Africa. But this is not so in respect of ethnicity and coups. Globally we can find no effect of ethnic polarization or diversity on the risk of a coup. But in Africa ethnic polarization sharply increases risk.

What else determines the risk of a coup? Well, the economy is important, just as in the risk of civil war. Coups are more likely the poorer the country and the lower its growth rate. So if the president adopts policies that promote the development of the economy, he becomes safer. Perhaps the president musters a flicker of interest at this point, or perhaps his eyes glaze over: not another homily on good economic policy. A further economic effect is via aid. After allowing for the possibility of reverse causality, an additional 4 percentage points of GDP of aid increases the risk of a coup by around a third. This may be because aid works like a honey pot, making control of the government more attractive. So, inadvertently, donors may be exposing governments to an enhanced risk of a coup.

What else did I find that might cheer a sleepless president? Ah, yes, coups are getting less common with the passage of time: they are gradually going out of fashion. The president concludes that all he has to do is tighten the repression and hang on: time is on his side. Unfortunately for presidents, this is in part offset by a countereffect. Each year that a leader stays in power increases the risk of a coup: far from gradually becoming indispensable, political leaders who stay in power for decades overstay their welcome.

For an incumbent president the passage of time and the length of incumbency offset each other. In any particular year, say 2008, a president who has been in power a long time faces a higher risk of a coup than one who is a newcomer. President Mugabe, who has been in power for twenty-eight years, is more likely to find himself past his sell-by date than President Mwanawasa of Zambia, who came to power much more recently. Similarly, we can compare the coup risks faced by two equally long-serving presidents at different times. The year 2008 is President Mugabe's twenty-eighth year in power. For President Eyadéma of Togo the same long-service milestone of twenty-eight years was reached back in 1995. At that time such long service implied that a president was living dangerously, although in the event Eyadéma reigned on and on until gathered up from the presidential palace by the Grim Reaper. Now, thanks to the passage of time, Mugabe is safer in 2008 than was Eyadéma in 1995.

What, apart from repression and economic development, can a president do to guard against a coup? One strategy that is widely touted in the literature is to divide the military into many different units so that each one can function as a check on the others. During the Kenyan attempted coup of 1982 the government was saved because the air force fell out with the army. In Zaire, President Mobutu split his military into so many units that were not allowed to communicate with one another that coups were made extremely difficult. He did, however, pay a price for that strategy since the same process made his security forces completely ineffective: despite its massive size, Zaire was unable to defend itself from an invasion by its tiny neighbor Rwanda.

Unfortunately, there are very few data on the internal structure of military forces, and so, while the divide-and-rule hypothesis sounds eminently sensible, it is very difficult to test. We hit upon one possibility: since landlocked countries did not have navies, all other things equal, their military was likely to be less divided, and so a coup attempt was more likely to succeed. We investigated whether this was borne out empirically. Although we indeed found that coup attempts were more likely to succeed in landlocked countries, the effect was not statistically significant, so it may well be pure chance. However, since the sample size for this test was small by the standards of statistical testing, little can be concluded from the lack of significance. My guess is that divide-and-rule works. The president is getting impatient reading all this: he has already divided his military into seven distinct groups, each headed by a cousin.

So let me try to be more helpful. I have found something simple and within the power of any president: adopt a term limit. At the start of the 1990s term limits became fashionable. If an incumbent president agreed to a limit of, say, two four-year terms, but declined to make the rule retrospective so that the clock only then started ticking, he had the prospect of a further eight years in power, and that seemed a long time. The adoption of term limits significantly and substantially reduces the risk of a coup: in fact it more than halves the risk. Armed with term limits, the incumbent presidents of the 1990s were a whole lot safer than they had been previously. But as the fateful year approached when the term limit was actually supposed to preclude them from continued power, presidents began to have second thoughts. Should they really step down? Perhaps this would be irresponsible? Surely they were indispensable? So, with heavy hearts, they bowed to the pressure from their sycophants who were themselves scared witless at the prospect of losing their access to patronage. They began the process of changing the constitution so as to remove the term limit.

The degree of difficulty that presidents faced in abolishing term limits was, in fact, a good measure of how vigorously the society had built constitutional defenses. The presidents of Chad, Zimbabwe, and Uganda succeeded in abolishing them; the president of Russia found an ingenious way around them by shifting to become prime minister. The presidents of Zambia, Nigeria, and Venezuela tried to abolish them but failed.

The evidence that term limits are effective in reducing the risk of a coup is the most encouraging result so far: it suggests that to some extent coups do function as last-resort checks on power. But whether term limits will continue to be so effective in reducing the risk of a coup depends upon their credibility. With so many presidents waiting until near the end of their final term and then removing the limit, those restive for power must now ask themselves whether a term limit is merely a trick. For those presidents who fooled their army into believing that they had committed themselves to an end-point only to remove it, the adoption of a term limit actually had the perverse effect of lengthening their expected period of office rather than shortening it.

To focus the mind, ask yourself what President Mugabe would make of all this. He would note that the collapse of the Zimbabwean economy has seriously exposed him to the risk of a coup. He would note that this is compounded by his having been in office for twenty-eight years, a mere blink of the eye compared with presidents Castro and Ghadafi, but nevertheless rather a long time. As to term limits, forget them. The only hopeful bit is that repression works. But the army is restive. What else could a worried president do?

The president controls the size of the military budget. If he is worried about the possibility of a military coup, he could change the budget. But life is complicated: in which direction should the president change it? He is, it appears, on the horns of a dilemma. If the army is the menace, then perhaps the safest thing is to slim it down. If each officer is a potential Napoleon, then the fewer officers, the safer the president. But offsetting this, if the army is demanding more money, then perhaps the safest thing is to pay up. The president dithers: up or down, which is best? At moments like this he has learned to turn to the Internet, and sure enough, he swiftly finds some research that provides the answer. At first the answer seems to be as hedged around as a Delphic oracle, but he sorts it out.

In most countries for most of the time the risk of a coup is negligible. If the head of the army comes along and starts muttering that the troops are restive, then a sensible president tells the army what he thinks of it. The chance of a coup being successful is so low that no sane army officer would try it: the penalties for failure outweigh the payoff to success because failure is so likely. So the extortion threat is not credible. The president points to the example of Costa Rica, which has managed perfectly well without an army, and cuts the budget. We find something like this in the data. In the normal range of coup risk, the level of military spending does not affect the risk, and governments respond to small increases in risk by cutting the military budget: if the military is a nuisance, you might as well have less of it.

But there is a different range of coup risk. If the risk is high, then the extortion threat becomes all too credible. The president knows that a coup would have a sufficiently high chance of success that the payoff might well be worth the risk. Unless he pays up he is living dangerously. If, however, he pays up and does what the army wants, then the payoff to a coup is reduced: he is safer. We also find this in the data. In the range in which coup risk is high, a high level of military spending significantly reduces the risk of a coup, and, consistent with this, in response to a high risk the government increases the military budget. I think of this as grand extortion: the army is menacing the government for money in much the same way as a gang of thugs run a protection racket, except that this is on a grander scale.

So, to interpret the Delphic oracle of economic research, all the president needs to work out is whether he is facing a high risk of a coup or a low risk. If the risk is low, he can do what he would sometimes rather like to do and slash the military budget, showing all those useless officers with their gold braid what he really thinks of them. If, however, the risk is high, then he had better raise the military budget. He will have to steel himself to face down the disapproval of the donors as he raids the health budget and announces a pay increase for the army. President Mugabe is in no doubt. The meltdown of the economy, his long period of incumbency, the absence of credible term limits: each of these raises the risk of a coup. He is in deep trouble. Repression plus money for the army seems to be his best way out. As a preliminary measure he doubles the police force.

IT IS TIME TO RETURN to that awkward-looking question that I parked: do coups typically lead to improvement or deterioration? We know that they do not come cheap, but if they are the only way of removing a bad regime, then perhaps they are to be welcomed.

There are two aspects of the legacy of coups that we might reasonably judge them by: their political consequences and their policy consequences. We decided to use the standard measures for each of them. For the political regime we used the score of the Polity IV index, and for economic policy we used the rating of the World Bank called the Country Policy and Institutional Assessment. Each of these has limitations, but they are a reasonable guide to the legacy of a coup. While the immediate impact of a coup might well be adverse, to judge its legacy you need to look longer term. We decided to look at the five years following a coup, year by year. We confined ourselves to successful coups: only these produce regime change. Before I describe what we found, think for a moment what the effects of a benign surgical-strike coup should look like. Quite possibly, even though the coup is benign, in the first couple of years outcomes might further deteriorate. But thereafter they should rapidly improve. We might reasonably hope that five years after the regime change, both the polity and economic policies should be significantly improved.

Such hopes would not be justified. In the first couple of years following a coup, the political regime does indeed significantly deteriorate. But even after five years it is still worse than before the coup. The story is similar for economic policy. For the first three years after a coup there is a sharp and significant deterioration, and even by the fifth year policy is worse than prior to the coup. Indeed, if you think back you may recall that one legacy of a successful coup was a sharp increase in military spending. Not only do presidents increase the military budget to try to ward off a coup, but if one succeeds, the ringleaders reward the army by slamming up spending. No wonder the World Bank assessment of economic policy deteriorates. There is one further legacy of a coup: it significantly increases the risk of a civil war. So the political legacy of coups is not particularly impressive.

There remains one possibility that I must confess I have not investigated. It may be that although actual coups are detrimental, the fear of a coup keeps politicians on their toes, forcing them to deliver reasonable policies. I have not investigated it because it is an extremely difficult proposition to test. Quantitative analysis depends upon differences: differences in the risk of a coup would have to show up as differences in government performance. I have not got a sufficiently strong stomach to try: there are too many obvious ways in which causality could flow in the other direction, with differences in government performance affecting the chances of a coup. I doubt that it is possible to find a test that would be convincing, and there is considerable scope for results that are misleading. However, accountability to the army need not improve government performance for ordinary citizens: it may well improve government performance for the army at the expense of ordinary citizens.

Evidently, if things are sufficiently desperate, a coup is to be welcomed. A coup is sometimes the only bloodless way of deposing a disastrous and illegitimate regime, and in such circumstances military officers do have a responsibility to take action. The alternatives may come down to popular protest and rebellion. Recall that popular protest against autocracy becomes increasingly pronounced only as incomes rise. At the very low levels of income that characterize the bottom billion, protests are relatively rare and readily squashed. Rebellions are too costly and unreliable to be a worthwhile route to political change. So coups have a role to play in maintaining decent governance, and the fact that they are getting less common is not necessarily good news. Yet the historical record is not encouraging. Surgical strikes do sometimes happen, but more commonly coup leaders are not surgeons wielding a scalpel, but rank amateurs hacking away at the body politic. To date, coups have been unguided missiles that have usually hit the wrong target. Rather than be eliminated, perhaps they need a guidance system.

Chapter 7

MELTDOWN IN COTE D'IVOIRE

Cote d'Ivoire brings it all together in one disastrous meltdown: a fraudulent election, a coup, another coup, and a war. Yet it used to be known as the African miracle. Its capital, Abidjan, was regarded as Africa's Paris.

To make sense of the meltdown we need to start with the prior success: what was the Ivorian miracle? Success had not been based on democracy, but on the vision of an autocrat, Félix Houphouët-Boigny. As you will see, his strategy was risky, but it nearly came off. Along the way, the president transferred the capital to his home village, Yamoussoukro, built an astonishing basilica modeled on St. Peter's, and induced the pope to come and open it. Since the basilica was financed partly by the diversion of aid, it was viewed with something between horror and derision by the donors. But societies throughout history have used monumental buildings to construct a shared identity. The anthropologist Colin Renfrew suggests that Stonehenge was such an edifice, and as I will argue, the creation of a sense of shared identity is very much what leaders should be doing in these societies. Whether a cathedral in the president's home village was the ideal symbol in a society divided by religion and ethnicity might, however, be questioned.

The high-risk strategy nearly came off, but not quite: Cote d'Ivoire is now regarded as one of the region's least tractable development disasters. Its transformation is a story of economic shocks, elections, guns, wars, and coups. The meltdown started with a mismanaged economic shock, was compounded by an election, followed by a coup, which then escalated into a war, fueled by a scramble for arms that induced the international community to impose an arms embargo, which failed. Indeed, Cote d'Ivoire during a single decade has all the events that this book is about. In what follows I have relied heavily on the expertise of Jennifer Widner, a political scientist at Princeton.

From independence until 1980, Cote d'Ivoire was a huge success. Houphouët-Boigny aspired to build a strong economy through a 1950s-style French model: strong state institutions supporting private-sector growth. This strategy contrasted markedly with the prevailing model of socialism. Indeed, at independence, the president of neighboring Ghana, Kwame Nkrumah, had challenged him to a wager that in ten years Ghana, with a standard socialist model, would far surpass Cote d'Ivoire. Nkrumah lost: by the 1970s Ghana was in a state of economic and political collapse, and he himself had been ousted by a coup, whereas Cote d'Ivoire was stable and prosperous.

The core of the growth strategy was immigration: immigrants were welcomed to come and cultivate cocoa on unused land. This produced a tidal wave of immigration from Burkina Faso, the landlocked, resource-scarce neighbor. By the 1980s an astounding 40 percent of the labor force was immigrant. Politically, the model worked because Houphouët-Boigny gave immigrants some political power and naturally won immigrant support. The quid pro quo for native Ivorians was that cocoa was heavily taxed. The revenue financed jobs in the civil service, and these went overwhelmingly to locals. Potentially, the longer this system continued, the more stable it would become: immigrants would become such a large bloc that they would be essential.

Since Houphouët-Boigny ran a one-party state, it looked as if he could afford to play it long, but in the event, his risky strategy was derailed by economic shocks. In 1980 the world price of cocoa and coffee collapsed and the price of imported oil rocketed. The ensuing economic crisis was partly met by borrowing: by 1993 debt had accumulated to $15 billion. Even with this massive borrowing, average incomes duly collapsed by around a third. Poverty soared.

The politics compounded these economic problems. The tax on cocoa had been disguised as a price stabilization scheme: the price was guaranteed, but at a level that had been below the world price. As world cocoa prices fell to levels nobody had anticipated, the price guarantee duly kicked in: the cocoa-producing immigrants were being subsidized instead of taxed! To keep the political deal in place the civil service continued to expand, exacerbating the collapse of the private economy. Whereas in 1980 half the urban workforce had proper jobs, by the early 1990s three in four were scratching a living informally: the urban poor were set to be a powerful political force. As jobs dried up and incomes fell, young men were forced to consider working the land. But by now the best land had been occupied by immigrants.

By the early 1990s Houphouët-Boigny was well past any reasonable sell-by date: he was an old man who had been in power for more than thirty years. But he was tenacious for power. To maintain his grip he created a highly confused situation concerning the succession. Then, in December 1993, he dropped dead. He had constructed confusion so brilliantly that he had become genuinely indispensable: his death was not announced for at least a week because rival contenders were battling it out. With no clear rules, whoever got the crown was going to face continuing challenges. It was inevitable that some political aspirant would exploit the potential for anti-immigrant sentiment. In the event, they all did so. Since by now the economy was a disaster there was a desperate need for economic reform, but any payoff was already mortgaged to repay debt.

The politics went wrong in discrete steps. Henri Konan Bédié became the president, but Alassane Ouattara, who had been the prime minister, had the stronger economic credentials. There was immediately a massive economic shock: the common West African currency was devalued by 50 percent, creating a powerful redistribution of income. The big losers were civil servants, whose wages were eroded. Since the civil servants had less to spend, those scratching a living in informal urban activities were also hit. In any devaluation the big beneficiaries are exporters: in Cote d'Ivoire this meant the immigrant cocoa farmers. The devaluation also launched an aid boom, essentially as a reward. Aid suddenly spiked from around 7 percent of income to more than 20 percent, and growth at last started to pick up. The new Bédié government thus started with both an opportunity and a crisis. The opportunity was economic recovery, but the crisis was a precarious mandate and a political powder keg of anti-immigrant sentiment. The very policy that opened the growth possibilities radically accentuated the political problem.

Bédié had beaten his ambitious and technocratic rival, Ouattara. Unfortunately, they did not manage to achieve the harmonious relations that in Britain Tony Blair achieved with the gracious personality of his defeated technocratic rival, Gordon Brown. Instead, after four months Ouattara left to a top job at the IMF and became the prince over the water. Since Ouattara was from the north and Bédié from the more populous center of the country, Bédié decided to play heavily upon the politics of identity. However, the first politician to play the anti-immigrant card was Gbagbo, a minor politician from an area where immigrants occupied much of the cocoa-growing land. Bédié followed suit, reversing the ruling party's political position by 180 degrees. One big advantage of the reversal was that Ouattara, being a northern Muslim, could be cast as an immigrant and declared to be a noncitizen. To make sure, Bédié changed the constitution to disenfranchise Ouattara as a candidate from the next elections.

As the 1995 polls approached, it was evident that no opposition politician other than Ouattara could command a significant share of the vote. Gbagbo avoided ignominious defeat by declining to stand and persuaded the gullible Ouattara-linked party to join him in boycotting the polls. Both opposition parties formed militias to enforce the vote boycott: recall that electoral violence tends to be the strategy of the opposition. By default, Bédié won in an election widely perceived to be unfair. By emphasizing identity, Bédié ignited a powder keg. Antipathy toward immigrants intensified as the political press broadcast highly inflammatory reports of unfairness of one community toward another. The president forged ahead with a policy to remove many northerners from positions in government.

Economic retrenchment and Bédié's pursuit of identity politics conspired to irritate the military. In the country's heyday the security forces were well paid but small: following the standard precautionary arrangements, it was divided into several branches: there were eight thousand gendarmes, sixty-eight hundred soldiers, eleven hundred in the presidential guard, nine hundred in the navy, and seven hundred in the air force. Even in the Houphouët era there had been a few coup attempts. In 1990 army troops had seized the airport outside Abidjan and mutinied to secure better pay. A general named Gueï had intervened and negotiated an end to the mutiny, being rewarded with promotion to chief of staff.

Gueï initially continued as chief of staff after Bédié's accession to the presidency, but there was little trust between them: the armed forces were disproportionately drawn from communities outside Bédié's ethnic base. Gueï refused requests from Bédié to arrest Ouattara and to put down electoral violence in Abidjan. Bédié naturally got scared of the army. He was in a difficult position: recall the dilemma, reduce it, or buy it off? He decided to reduce it, but gradually, by salami slicing. He started by dismissing General Gueï along with seven hundred soldiers.

What of the prince over the water? Objectively, Ouattara's best

prospects of ousting Bédié were if the economy continued to unravel. Once reforms were launched, the fate of the economy was
seen as being in the hands of the IMF. Astoundingly, Ouattara was
now number three in the IMF, creating an acute perceived conflict
of interest.

Economic reforms were massively redistributing income to
immigrants at a time when immigrants were inevitably hugely
unpopular. Politicians were bound to play the opportunistic anti-
immigrant card. The regime's reformers were boxed in by the political priority of Bédié, which was to weaken Ouattara. Frustrated
by the slow pace of reform, the IMF, the French Treasury, and the
World Bank all came to the view that Ouattara was the solution
to the problem: aid was rapidly curtailed. Within the government
there was an understandable perception that these institutions were
playing for regime change.

I recall in late 1999 speaking in Cote d'Ivoire at one of the
surreal occasions that the development agencies love to sponsor: a
conference on good governance. With sublime incongruity it was
presided over by President Bédié. Sure enough, it did not take long
for governance to get decidedly worse.

Bédié's manipulations to maintain power generated the second discrete step in the move toward civil war: a military coup. On
Christmas Eve, 1999, about 750 Ivorian troops mutinied over unpaid
bonuses. A group of senior military officers went to see President
Bédié to demand higher spending on the army. He fobbed them off,
telling them to come back the following week. Instead they came
back later that day and deposed him. Whether General Gueï was
behind the coup from the start or brought in to legitimize an otherwise desperate situation is unclear. In any event, the former general
assumed control and transformed the mutiny into a bloodless coup.
Gueï promised fresh elections within six months.

Recall the French security guarantee for Francophone Africa.
Prior to Rwanda, any coup attempt in Cote d'Ivoire that had got out

of hand would have been put down by French troops. But on this occasion the French chose not to intervene. Gueï convincingly posed as offering a neutral brief interlude of clean-up: this was the surgical strike. Perhaps for a few weeks he even meant it. From this point on a rapid political sequence took the society to civil war.

Once Gueï was in power everything rapidly began to unravel. He indeed stuck to the commitment to hold elections within six months. However, once in power he realized that he had a natural proclivity for the job of president that it would be wrong to deny the nation. So he decided that he himself should be a candidate. From Gueï's perspective, however, elections posed a problem. Although his talents were evident to himself, there was no great upwelling of voter support: the country was polarized between those who wanted Ouattara to be president and those who wanted Bédié back. Fortunately, Bédié himself had demonstrated how to deal with such a difficulty. Gueï declared them both to be ineligible, securing agreement from his handpicked Supreme Court, which, having got the hang of it, also ruled out a further twelve candidates.

Had he looked to the role model of President General Abacha of Nigeria he might have been spared his one blunder: Abacha had pioneered a multiple-party election in which each of the five parties chose Abacha as its candidate. Sadly, Abacha had died before being able to contest his planned election against himself. Being less imaginative, Gueï decided that he ought to have an opponent. He accepted the kind offer of Laurent Gbagbo, the sure loser, to run against him so as to legitimize his victory. In doing so Gueï made the classic error of dictators, an overestimation of how much his people loved him. Most people did not bother to vote in this sham election, but among those who did, most voted for Gbagbo.

Normally even this inconvenience would not have derailed an incumbent president, let alone one who ran the army. The purpose of an election was to anoint the incumbent with the magic oil of democracy, not to choose a president. Sure enough, Gueï simply declared himself the winner and disbanded the electoral commission. Evidently we should regard this as a further coup.

However, Gueï's truly serious miscalculation was to overestimate not his votes but his firepower. Gbagbo had massively expanded his armed militia, the Young Patriots, financed by President Compaoré of Burkina Faso, who was annoyed by the xenophobic policies that Gueï, in a policy turnaround, had swiftly embraced. In response to the coup attempt, Gbagbo deployed his militia of violent and disaffected youth against the army. Normally a gang of youths versus a professional army would not stand much chance, but the army had been purely decorative and very small: it had never expected to fight and was not prepared to do so. It was also divided: indeed, some of its officers had already attempted a further coup against him. Gueï had responded by gutting the army that Bédié had already been salami slicing. As a result, within the narrow confines of central Abidjan the militias were able to outfight the army. They also turned on northerners living in the capital, dumping bodies of those they killed in the lagoon. Gbagbo came to power through the mixture of an illegitimate election and a rebel uprising.

Under the circumstances it might have been reasonable to restage the elections, as both Bédié and Ouattara duly requested. However, since Gbagbo would have been heavily defeated against either of the major politicians, he had no interest in holding a fair election. He used his party connections with the French socialist government, which duly recognized his victory. As president, his continued survival in power *depended* upon avoiding a further election. This in turn depended on the situation becoming and remaining sufficiently perturbed that elections could not be held. In 2001 there was the first of thirteen internationally brokered efforts at reconciliation, all of which failed.

Having managed to lose an election even against his handpicked opponent, Gueï had only one route back to power, for which he had evidently acquired a taste. Sure enough, in September 2002 he staged a comeback coup. Several hundred soldiers participated in attacks in Abidjan, Bouaké, and Korhogo, with Gueï in charge of the Abidjan rebellion. In Abidjan the coup failed: again, within the confines of street warfare, Gueï's army was no match for Gbagbo's militia, and Gueï himself was killed along with his family. The rebellious soldiers retreated north to Bouaké and Korhogo.

Within a week of the failed coup attempt, the soldiers who had rallied around Gueï were joined by an array of excluded politicians. They quickly seized towns in the north and center of the country, calling themselves the Forces Nouvelles (FN). The third, failed coup attempt thus evolved into a rebellion and hence into civil war.

Outside urban areas a conventional army with heavy equipment can easily defeat a youth militia, and so the FN advanced rapidly on Abidjan. At this point President Gbagbo had few options. If he stayed put and fought it out his fate would be the same as Gueï's. If he went into exile and appealed to the international community, the end result would be an internationally supervised election, which he would most surely lose. Gbabo's only card was his capacity for violence within Abidjan: could this be useful? He could use the young street gangs to murder some more northerners, but where would that get him? Was there any group in Abidjan whose vulnerability to violence might be turned to advantage? Recall that until the coup Abidjan had been Africa's Paris. This description was not entirely figurative: it had by far the largest concentration of French citizens in Africa. Gbagbo used them as hostages, demanding that the French army come to his defense. They arrived within three days to reinforce Gbagbo's position: French troops had to defend his regime to avoid a massacre of French civilians. This accounts for the extraordinary spectacle of Gbagbo denouncing the French to mass rallies of his youth supporters and indeed inciting youth to kill French civilians in Abidjan at the same time as the French army was defending his regime against the FN.

The French military imposed a cease-fire line, Operation

Licorne, and forced the FN to withdraw around one hundred ki-
lometers from positions that they had won. This created a strong
signal that the French government was not impartial. At this stage,
outside of urban areas, the FN was in a position to advance on the
south, and with its finance at least partly dependent on unsustainable outside sources in Burkina Faso, it was in a hurry.

The international community again tried to broker a power-
sharing agreement during negotiations in Paris. The agreement set
up a coalition government under Gbagbo, but assigned some of the
most important cabinet positions to the FN. Specifically, the rebels
would hold the post of minister of defense. This was the guarantee for the rebels of post-settlement security. Once the peace deal
was taken back to President Gbagbo for ratification he rejected this
component of it, whereupon it fell apart. In the process he inadvertently signaled that he had every intention of reneging on the peace
deal.

Gbagbo appears to have realized that time was on his side. Although initially weaker militarily than the FN, he had the larger
revenue stream and invested in buying armaments. The United Nations and the regional organization, the Economic Community of
West African States (ECOWAS), responded by placing an embargo
on exporting arms to Cote d'Ivoire. However, this did not prevent
both sides from acquiring guns. Weapons from Belarus and other
poor-governance countries came into the country via Togo. Gbagbo
even acquired an air force. His forces began to move against the
north in violation of the Paris agreement. The French peacekeepers,
whom Gbagbo had previously needed for protection, were now in
the way of his own attack. He therefore ordered his new air force
to bomb the French base near Bouaké, killing nine soldiers. The
French retaliated by destroying his air force.

The conflict attracted neighbors and predators. Mercenar-
ies from Liberia and Sierra Leone, hired on a "pay yourself" basis,
preyed upon Ivorian citizens and were responsible for some of the

bloodiest attacks. In addition to the major armed groups there were at least nine unofficial militias: organized violence was all too feasible in this environment. The methods all parties used to finance the conflict drew in unscrupulous companies, countries, and leaders. The Central Bank of West Africa was robbed first in Abidjan and then in Korhogo. Money came in from the neighbors: both President Compaoré of Burkina Faso and President Taylor of Liberia were generous. The FN created an "economic police force" to patrol diamond areas and levy taxes.

From this situation it was going to be difficult to reach a settlement. The French cordon of troops kept the fighting limited so the cost of the conflict was not so severe as to force the parties to negotiate. Since neither side had the slightest trust for the other, there would normally have been an important role for the international community to negotiate a settlement. The attempts failed because the only type of settlement that the international community could endorse was one that was validated by free and fair elections. But such elections would inevitably hand power to one or other of the serious Ivorian politicians, Bédié and Outtara, both in exile. Worse, from the perspective of the FN in control in the north and Gbagbo in control in the south, Bédié and Outtara managed to patch up their differences sufficiently to form a common electoral front: they would fight the election under one party. So, from the perspective of both the incumbent leaders, an international peace settlement was equivalent to defeat. The only hold that the international community had over them was that Gbagbo's term as an elected president expired. With his term expired, Gbagbo's legitimacy began to look highly suspect. In a deal forced on him by the international community, his own government was dismissed and a neutral technocrat brought in as prime minister, a change described by some members of his government as a coup. It was hard to envisage how this stalemate could be unblocked.

And then a settlement appeared out of nowhere: certainly without the participation of the international community. It was an internal settlement, between Gbagbo and Soro. Gbagbo got rid of the technocrat and appointed the rebel leader, Soro, in his place. Bédié and Outtara were excluded. Gbagbo and Soro promised elections in due course, but now there was no international community to require that excluded candidates could stand, or that the elections would be free and fair. As you have seen, under such conditions incumbents have a range of strategies for winning an election, and so at last Gbagbo and Soro need not regard peace as the path to the hell of electoral defeat. The internal settlement was potentially very attractive for both of them. Aid could be resumed, and Cote d'Ivoire's offshore oil could be tapped without awkward questions. The politics was brilliant. Within a month Soro narrowly escaped death in a helicopter accident.

Cote d'Ivoire is at last back to peace. But a decade of coups, war, and elections have taken their toll. The mantle of Francophone Africa's flagship has passed, largely by default, to Senegal. Could anything have been done to avert this catastrophe? It is time for some solutions.

Part III

CHANGING REALITY: ACCOUNTABILITY AND SECURITY

Chapter 8

STATE BUILDING AND NATION BUILDING

F AMOUSLY, PRESIDENT BUSH BEGAN BY deriding state building and ended up attempting it. I am going to suggest why it is so difficult. The now-successful states were built through a painfully slow and circuitous process of formation that turned them into nations with which their citizens identified. This enabled them to undertake the collective action that is vital for the provision of public goods. In the high-income societies we have come to take these features for granted: so much so that we have forgotten that they are essential. Legally, states can be built by the stroke of an international pen: they need only recognition. This is how the states of the bottom billion came into existence. They have not been forged into nations, and so they face an acute lack of public goods.

Most modern states were once ethnically diverse. The boundaries of a modern state generally emerged not out of deepening bonds forged out of a primordial ethnic solidarity but as the solution to the central security issue of what size of territory was best suited to the creation of a monopoly over the means of violence. Often the sense of a common ethnic origin bonded to the national soil was imagined retrospectively: conjured up by the urban, middle-class, romantic nationalists of the nineteenth century.

State formation was driven not by a sense of community but by the unusual economic properties of violence. We now know that violence is not something that emerged as a result of the formation of states: on the contrary, stateless societies are horribly violent. The production of violence depends upon the available technology. Hunter-gatherer societies are inherently extremely violent because the technology does not permit anything else: the winning strategy for a group of hunter-gatherers is the preemptive strike against neighbors through the predawn raid, catching your enemies detached from their weapons. Any group sufficiently quixotic to trust a peace deal gets eliminated before it can change its mind. So violence is intrinsic to such societies: they would more accurately be described as hunters, gatherers, and killers. However, with technological advance the production of violence becomes subject to specialization and economies of scale. Both make violence a paying proposition.

Start from a primitive landscape with no government and many identical households and now introduce a minimal degree of differentiation. Some people are more productive than others and some people are stronger than others. From the resulting four different types of people, ask yourself how one type, the unproductive strong, are going to earn a living? They are going to plunder those who are productive but weak. By abandoning their incompetent efforts to produce and specializing in violence, the unproductive strong get even better at violence. Violence requires skill and hence gives an advantage to professionals.

Onto this scene of specialization now add economies of scale in violence, this being a fancy way of saying that size matters. It is this that makes violence distinctive. Other economic activities had to wait until the industrial revolution before scale became important. A thousand-person farm was no more productive per person than a one-person farm; a thousand-person firm of cobblers was no more productive per person than a solo cobbler. But a thousand-person army could kill, one by one, a thousand solo fighters: large groups of professionals tend to defeat small groups of professionals. Not always and everywhere: small armies can win if they have better technology and better management; there is even room for differential heroism. The race is not always for the swift, but that is where to put your money.

So, by forming or joining a large group of professionals that establishes a monopoly of violence over a territory, you as a member become safer from attack. That is clearly a powerful incentive. But safety is not the only consideration: life can only be sustained with income. People specialized in violence forgo the chance to produce. Where is your income to come from? The answer, as any mafioso knows, is that having established a monopoly of violence, you now have the power to extort from other inhabitants of the territory. Why do the inhabitants not run away? Perhaps your army can enforce penalties for attempting to escape: you are able to turn the inhabitants into serfs. Perhaps the inhabitants have nowhere to run because the neighboring territories are dominated by similar armies, so flight would merely get them out of your frying pan into some other army's fire. Perhaps the protection from other predators that is a consequence of your local monopoly of violence is worth the payments. You, the army, are inadvertently supplying a public good: you have become a state.

Although the public good of security for the locality may be inadvertent, you gradually realize that it is in your own interest to supply a few other public goods. One is to help your inhabitants to trade with one another. If they become richer, then you can become richer by taxing them. So you provide a contract-enforcement service for them; after all, you are good at enforcement. You call it a court, and around it grows a legal system. You might also run to some trade-enhancing infrastructure: roads, bridges, and market-places. You might even, though this takes a certain amount of vision, put a few limits on yourself. By closing off some options, you make your richer subjects less inclined to adopt the infuriating defensive strategy of refusing to invest. We have arrived at a state, but not a modern one: the range of public goods is too limited because the interests of many people are ignored.

The final step from a state that is effective but serves the interests of a minority to one that serves everyone is another long haul. Once hemmed in by neighboring states, these become the primary threat: either you defeat and swallow them or they defeat and swallow you. Arms races develop. This requires high taxation, and the warfare generates a sense of nationalism: people start to sense a common identity. As the effective state facilitates economic growth, even the politically weak become better off, and this, together with an emerging sense of common identity, gradually makes them more powerful. Recall that autocracies become more prone to political violence as income rises. More specifically, they become increasingly beset by riots, demonstrations, and political strikes. The sense of common identity further eases the collective action of protest. Better provision of public goods is gradually prised out of the elite by this pressure. To make these improvements credibly permanent, elites also concede limited extensions of the franchise: the society inches toward modern democracy.

IN TRYING TO APPLY THESE simple but powerful economics of violence to the actual history of state formation, it is always convenient if we can find a starting point for history. In the process of European state formation, to my mind, the natural starting point is the fall of the Roman Empire during the fifth century. This has some rudimentary analogy to the decolonization of Africa in the mid-twentieth century. Given the suddenness of the decolonization of Africa, which was basically over a decade after it had first been seriously contemplated, the closest analogy is with the decolonization of Roman Britain.

The decolonization of Roman Britain was even more abrupt than that of Africa. Britain had the single largest unit of the Roman army, around 15 percent of the imperial force, paid for by astonishingly heavy taxation of the British economy. So whoever was in charge of this army had the potential for conducting a coup d'état. As the Roman Empire hit political turbulence in the late fourth century, twice in twenty-five years the head of the Roman army in Britain tried to become emperor. Since the first attempt in 380 had failed, the leader of the second attempt in 403 decided to improve his chances by taking his army with him on a march toward Rome. He still lost, but in the process Britain suddenly lost its army. Since the Roman government in Britain had been military, not only did Britain lose its army, it lost its government. The history of Britain post-403 makes the post-colonial history of Africa look like a staggering success. Within a few years the British had petitioned Rome to be recolonized: even heavy taxation was preferable to the absence of security and government. But Rome was not in a position to respond, so British society was left to its own devices. What followed was a descent into civil war, the collapse of public goods to the extreme extent that the urban economy disappeared. People fled the country, the mass of emigrants across the Channel nostalgically naming their new home Brittany.

So this is our beginning: post-Roman chaos. It took Britain, and indeed the rest of Europe, centuries before local thugs coalesced into miniature states, each able to keep a degree of order within its own territory but fearful of its neighbors. By 1555 the German-speaking territories still had no fewer than 360 states. Gradually the states became more frightened of one another than of threats from within their own societies. To defend against neighbors they needed a large standing army. Big defense costs money, and the only sources were taxation or borrowing on a scale not seen since the days of the Roman Empire. Taxation has its limits. If people are taxed beyond their willingness to pay, they will take evasive action, conniving with the tax collectors so that they bribe the collector instead of paying the state. Ultimately, if taxes get too onerous, people retreat into activities that cannot be taxed.

Borrowing is even more of a potential minefield for the state. Whereas taxation is basically coercive, borrowing depends upon people actually volunteering to lend the state their money. Even if they are prepared to lend it, they demand interest, and if the interest rate is high the borrowing becomes unsustainable, the military effort collapses, and the state is defeated.

The first European state to discover how to raise money on a sustainable basis through taxation and borrowing was the tiny commercial state of the Netherlands. This tiny society had a territory badly suited to defense: recall that mountains come in handy. The Netherlands is the least mountainous country in the world. Worse, its citizens were disproportionately urban and bourgeois, not groups with a strong fighting tradition. The Netherlands was facing a massive war machine: the Hapsburg Empire. In this David-and-Goliath struggle, David was sufficiently desperate that it had to evolve one advantage: the ability of the state to raise money. Even here it was up against a huge disadvantage: the Hapsburg Empire had the gold and silver mines of Spanish America.

The critical invention of the Dutch was political accountability. People were only prepared to tolerate high taxation if the government of the state became accountable to citizens. Not all citizens, of course, but the rich citizens who were paying the taxation. Further, with an accountable state the government was able to borrow: people were prepared to lend once they saw that the government was being forced to conduct its finances in such a way that it would always be able to pay them back. The Hapsburgs found that gold and silver were not quite enough, and so they too decided to borrow. But nobody had forced them into accountability. And so the battle for the Netherlands turned into a battle of interest rates. The power of compound interest to gradually gut the finances of a profligate borrower ensured that final victory would go to the state with the better credit rating. The Hapsburgs had a huge empire and the bullion mines of Spanish America as collateral, and the Dutch had a tiny area and political accountability. The power of compound interest takes time, but the Dutch were able to borrow for around 6 percent whereas the Hapsburgs were paying up to 22 percent. That is, when they could borrow at all: before the end of the war they had gone bankrupt and were shut out of the credit market. David beat Goliath.

Gradually, other states learned the Dutch lesson. Those that didn't got swallowed by those that did. Interstate warfare had two consequences. One was the sentiment of nationalism. It was to rationalize these sentiments that the educated, urban romantics of the nineteenth century conjured up the notion of deep ethnic roots that defined the nation. The clash of states became the clash of ethnicities: the myth of a common ethnic identity was forged on the battlefields. The sense of a common enemy and the myth of shared ancestral origins unified the inhabitants of the state into the people of a nation. The result was potent. As a benevolent force it provided the bonds that, via protest, enabled the ample provision of public goods: probably for the first time in history the collective action problem was overcome for the common good. As a malevolent force it generated vilification of the other: for example, in the First World War the British press was routinely describing Germans as Huns.

The other consequence of warfare was the spread of fiscal accountability: governments had to become accountable to the rich, otherwise they could not raise sufficient taxation and debt. But at this stage states still had not reached anything that looked remotely like the modern liberal state. It was not yet democracy and it was certainly not yet the use of taxation for social spending. The states of the mid-nineteenth century were run by the rich, and their priority was national security. The road from there to the present is paved with political protest from the excluded. Gradually, little by little, to avoid worse, the rich expanded the franchise. This enabled them credibly to commit to redistributive reforms that became irreversible without being so drastic that the economy was damaged. Nations inched toward democracy, and as they did so the priorities of government inched toward the priorities of ordinary citizens— the supply of public goods such as health and education instead of simply defense. Gradually the state became captured by the interests of ordinary citizens: we have arrived at the modern liberal democracy.

The evolution of the modern state was, on this analysis, violence driven. Step by step, the predatory ruler of the mini-state had evolved into the desperate-to-please, service-promising, modern vote-seeking politician. Such have been the crooked byways by which the modern state has evolved into its role of providing public goods.

Potentially, scale economies in violence permit the continued coalescence of states into superstates. The world has repeatedly seen the emergence of such enormous military territories: Rome, the Mongols, the Hapsburgs, the British, the French, the Portuguese, the Russians, and the Austro-Hungarians. Often the process is very rapid: technology can permit states to expand explosively. The development of the stirrup in the geographic context of the steppes suddenly enabled the Mongols to build the largest land empire ever known. Similar expansions occurred during the nineteenth century. When the pace of expansion gets sufficiently far ahead of the process of building a common identity, the resulting superstates face overwhelming problems in trying to establish a common identity. Instead of becoming *nations*, by default they become *empires*.

Nation building depends upon the choices made by political leaders. Their choices influence the pace with which empires turn into nations. The Romans took centuries but eventually began turning their empire into a nation by granting rights of citizenship to its inhabitants. At the other end of the spectrum of leadership incompetence, Haile Selassie was so besotted with the idea of being an emperor that within a decade he turned the new federal state of Ethiopia and Eritrea into an Ethiopian Empire with Eritrea as its colony. By the time he did this, his strategy was doomed: the age of empires was over.

The age of empires came to an abrupt end for a variety of reasons, but probably the most powerful was the rise of America to primacy and its resolute antipathy to them. The seeds were sown by President Wilson at the Paris Peace Conference after the First World War. Wilson committed himself to the principle of self-determination of peoples, a concept entirely revolutionary to the then-established principles of international relations. Self-determination implied that instead of identity continuing to adjust to political borders, borders would be adjusted to wherever identity formation had been reached: the music had stopped and peoples rushed to sit down on the chairs. Self-determination was put into practice in the Versailles Treaty, most notably in the territorial mosaic that in due course yielded the catastrophe of the Balkans, but it really came into its stride after the political showdown between America on the one side and Britain and France on the other that constituted the Suez crisis of 1956. Following Suez the British rapidly dismantled their empire, creating precedents that forced the French and Portuguese to follow. Ultimately self-determination even dissolved the Russian Empire. As a result, during the second half of the twentieth century the number of independent states increased massively.

This process of state formation was entirely different from state formation Mark I. With rare exceptions, the new states did not emerge as the solutions to struggles to provide security. It is usually said that the boundaries of the new states were arbitrary. This is not entirely fair to the colonial authorities that faced the task of turning a vast multitude of ethnic communities into manageable countries. The fundamental problem was that neither of the two processes that had happened in the formation of modern states had taken place:

there had been neither the emergence of territories viable in terms of security, nor the retrospective creation of an imagined community among the inhabitants of these security-defined spaces. In Africa alone there were some two thousand ethno-linguistic groups. Yet if each were made a nation, its territory and population would be far too small to reap adequate scale economies of security: they would be insecure both internally and externally.

Thus, although the instant states that came into being with the dissolution of the colonial empires were ancient societies with a multiplicity of strong ethnic loyalties, usually they lacked national loyalty: people's primary allegiance was to their ethnic group. As I have argued, this severely impeded the provision of public goods. Anything public was simply up for grabs: a common pool resource, the control of which depended upon winning the political struggle between the various ethnic groups. Much the surest way of overcoming this problem would be to follow the earlier model of nation building: gradually erode ethnic identities and replace them with a national identity.

One reason that ethnicity is considered an embarrassing topic by many Africans is that it is seen as a throwback, the antithesis of modernity. As modernization proceeds it will surely fade with time. This is a comforting proposition, but as is repeatedly the case, being comforting does not make a proposition true. There is no substitute for evidence. The evidence from recent surveys of attitudes across nine African countries by Afrobarometer is not encouraging. It found that if people are educated they are *more* likely to identify themselves through their ethnicity. The same is the case if they have a wage job as opposed to the traditional occupation of farmer. The same is the case if they have experienced political mobilization. So development, with the attendant education, jobs, and electoral competition, is increasing the salience of ethnic diversity rather than erasing it. Perhaps this is because it is in the modern economy rather than the traditional economy that the ethnic political contest is being played out. Farmers can stay semidetached from the consequences of ethnic politics, but if public sector jobs are assigned on the basis of ethnic allegiance, then education and wage employment would indeed make ethnicity more important.

Yet if the many disparate ethnic communities had been packaged together into a few states large enough to be secure, they would have faced a horrendous task of giving their inhabitants the emotional identity necessary for a state to function. In the event, the two thousand ethnic groups inhabiting Africa were bundled into fifty-four national territories. Was this too few states, resulting in unmanageable ethnic diversity, or too many, resulting in a lack of security economies of scale?

The decolonization of the bottom billion produced a patchwork of little states not utterly different from the situation of post-Roman Europe. But from then on the stories diverge. To a large extent borders of the bottom billion have been frozen: they did not face powerful challenges from their neighbors, at least not to the extent of fearing that they would be absorbed. I can think of only two mergers between countries in the past fifty years, both in 1989: the East German ambassador to North Yemen was uniquely unfortunate in becoming doubly redundant. The general trend has been the opposite, a further splitting of already small nations as rights of self-determination became recognized. And so, despite the arms races in Lilliput, the governments of the bottom billion have not engaged in international wars to anything like the same extent as did the European states of the nineteenth century. The resulting reduced need to tax has been reinforced by aid: in the typical country of the bottom billion the government gets around a third of its expenditure met by aid. The combination of modest military spending and high aid has left the tax burden quite light: often around 12 percent of GDP. This level of taxation has been too low to provoke citizens into demanding accountability.

I began to think more rigorously about how a corrupt ruler

might view taxation. Suppose, say, that you were President Mobutu, how heavily would you have taxed your society? It struck me that the lightness of the taxation may have been a deliberate strategy. Mobutu clearly wanted money, and he was also periodically pretty short of it. Mobutu did not amass a huge fortune; the revenues he grabbed from the Zairean state were used to buy loyalty from his enormous entourage. His first and foremost source of revenues had been to bleed the companies that were extracting natural resources. But once he had ravaged these companies to the point of ruin he did not turn to heavy general taxation, instead he turned to the printing press, the same solution that President Mugabe has hit on.

Hyperinflation is a very high-yielding form of taxation, and what is best about it is that people do not recognize it as a tax. In fact, it is a tax on holding money. If prices double every month, as they did at one stage in Zaire and are doing at present in Zimbabwe, then effectively the state is imposing a monthly tax of 50 percent on all the cash that people are holding. Work out what the state gets. Take a typical person who gets paid monthly and spends his income evenly through the month. On average he will be holding two weeks' worth of income as cash. So 50 percent inflation grabs one week's worth of income. Since it does this every month, over the year it amounts to a 25 percent tax on income. Not bad for a tax that people do not regard as a tax! The reason hyperinflation is not more common is that the revenues do not last. As people get used to high inflation they find ways of holding less money relative to what they spend: for example, they buy as much as possible as soon as they get paid. That is why hyperinflations are explosive and end in tears. Both Mobutu and Mugabe used it only as a last resort. As an addendum I will take the opportunity of final revisions to the text to update the figure on Zimbabwean inflation. Prices are no longer doubling every month: they are doubling every week.

Corrupt rulers might be wary of explicit taxation because of its capacity to provoke opposition. They do not want to tax so heavily that they provoke irresistible demands for accountability. It is no good having huge tax revenues if they then have to be spent on things that benefit everyone: your supporters will have no reason to stay loyal if they are rewarded no better than anyone else. So you must trade off high taxation against higher accountability. Economists like to set out choices as decision problems in which somebody is trying to maximize something: a firm might be trying to maximize profits, or an individual to maximize happiness. Indeed, crude as it is, this is what gives economics its enormous potency: we can work out what choices would be made if people were actually trying to maximize. Crucially, we can then work out how their choice would change if the world they are facing changes. Generally these predictions are not a bad approximation to reality, and that is what keeps economists in business.

I realized that the corrupt politician's choice could be set up as a simple decision problem: choose the tax rate that maximizes what you are free to embezzle. A very low rate is no good because there is no revenue to embezzle, and a very high rate is no good because although there is plenty of state revenue, it is defended against embezzlement by the scrutiny that the taxation has provoked. From the perspective of the corrupt leader there is an ideal rate of taxation, and it might well be quite low. We can also use this framework to infer how much beneficial public expenditure takes place under the rule of the corrupt leader: it is not zero. The corrupt leader would like it to be zero: from his perspective spending on what ordinary people want is a waste of public money that he would prefer to use for patronage. But, having set the tax rate at the level that maximizes patronage money, the leader has to live with whatever level of scrutiny that opposition to taxation has provoked. If, say, the level of scrutiny enables him to embezzle one-third of the state revenues, that still leaves two-thirds that are spent properly. The overall revenues are lower than they should be because the leader has kept taxation artificially low so as to depress scrutiny. So citizens are hit twice over: they only benefit from two-thirds of the revenue, and the level of revenue is lower than they would like. They still get some public goods, but this is not a sign that the ruler had some goodness in him after all.

This sketch of how accountability and a sense of nation evolve provides a rudimentary explanation for the political problems of the bottom billion: they are stuck. The state is ineffective partly because it would not be in the interests of leaders for it to be more effective, and partly because the supply of public goods is impaired by the lack of a sense of common identity. Based on the analogy with the formation of effective states in Europe, the solution would be greater state military rivalry. As states felt less secure against one another they would need to raise more taxation and this would provoke greater accountability. It would also presumably generate a strong sense of national identity.

I am going to argue that this is not an acceptable solution, but before we discard it I will set out a little evidence in its favor. Among the leaders of the bottom billion, President Museveni of Uganda has been unusually effective. When he came to power in 1986 the society was quite literally in ruins: it had taken less than a quarter century of independence to pass from peace and growing prosperity to mass violence and impoverishment. Uganda was, indeed, not a bad approximation to what Britain must have gone through after the Romans pulled out. Kampala, like fifth-century London, was reverting to the bush. President Museveni has achieved a remarkable transformation. Despite being landlocked and resource-scarce, Uganda has been one of the fastest-growing of Africa's economies. He has consistently placed the interests of economic recovery above the patronage and populism that have been so common elsewhere on the continent. What was the driving force behind him: what was his ambition as a leader?

I got to know President Museveni and I came to have great admiration and respect for him: I came to realize that he was not only a statesman but he was a military leader with ambitions for changing the political architecture of Eastern and Central Africa. For this he wanted a strong army. The man whom he most despised was his predecessor, President Amin. Amin had not only wrecked the Ugandan economy, he had suffered the ignominy of being deposed through an invasion by Tanzania, whose army had routed his own. One lesson that I believe President Museveni drew from this was that without a strong economy there could be no strong army. I think this was the bedrock that underpinned economic reform.

He not only rebuilt the economy, he conducted Africa's only truly successful campaign against AIDS. His leadership of this campaign, Zero Grazing, was decisive because it persuaded ordinary Ugandans to change their sexual behavior. Helen Epstein brilliantly describes it in her book *The Invisible Cure*. What she doesn't reveal is the key step in convincing Museveni to act. Given that his army was his priority, Museveni arranged with Fidel Castro that his officer corps should be sent to Cuba for training. Once in Cuba his officers were given medical checks. The message came back from Cuba: do you realize your officer corps is overwhelmingly HIV positive, they are going to die of AIDS? I suspect that Uganda's AIDS campaign, like its economic reforms, was in part motivated by President Museveni's military ambitions.

Uganda certainly has not gone all the way to being an accountable polity, but it is nevertheless a genuine example of increased state effectiveness. A similar story is Rwanda since 1994. The government of Paul Kagame, like President Museveni a successful rebel military leader, is currently the leading African example of effective state building. Museveni and Kagame jointly invaded and occupied Zaire, whose army had collapsed under the state-destroying patronage of Mobutu. They then fell out, and their mutual penchant for the military turned into an arms race against each other: I recall the outrage of Clare Short, at that time the secretary of state in charge of Britain's aid program, on receiving a letter from President Museveni justifying yet another increase in Uganda's military budget on the grounds that Kagame was plotting to invade Uganda. So there are two examples of military ambition and rivalry leading to state strengthening.

However, I balk at the notion that the societies of the bottom billion need to go through the same process as Europe. Even if the solution eventually worked, it would be at enormous cost. Europe tore itself apart with wars, and I do not wish to see the bottom billion do the same. War is even bloodier now than it was when Europe was fighting. There simply has to be a better way of building an effective and accountable state because the war route is utterly appalling. But I do not want to be guilty of believing something because it is so much more attractive than the alternative. Self-deluding thinking has bedeviled issues of development for decades. We have to work within the world as it is, rather than the world we would wish. So, while the appalling cost of the historical route is a good reason for *hoping* that there is a better alternative, it is not a good reason for *thinking* that there is one.

Soon I am going to set out my basis for believing that there is a better way. But first let me stay in destructive mode and explain why I think that even the historical route is no longer an option. If I am right in this but wrong in thinking that there is a better way, then the implication would be that the bottom billion would persist: there simply would not be a route to accountable and effective statehood. Some thoughtful people assert just that. Michael Clemens, writing in the highly influential journal *Foreign Affairs*, concluded that the bottom billion had no chance of development within our lifetime.

So why is the historical route now closed off? Partly because the manifestly high costs of international warfare and military rivalry make it politically unrealistic: neither the societies of the bottom billion, nor the international community, would let it happen. But, over and above these concerns, it would not work. Even if the bottom billion went through a long process of warfare against one another, they would not end up with effective and accountable states. The key reason is that many of the governments of the bottom billion now have huge revenues from natural resources. There are too many countries in the financial position of the Hapsburg Empire. They could inflate their military spending for many years on the back of their natural resource revenues without recourse to domestic taxation. Indeed, the government of the bottom billion that set its military spending at the highest level was Angola, which for a while was spending 20 percent of GDP on the military. Yet it had no domestic taxation and is one of the least accountable governments of the bottom billion.

So what are the realistic options? Surely the best is the route taken by President Nyerere in Tanzania: political leadership that builds a sense of national identity. Astonishingly, Nyerere achieved this without resorting to the notion of a neighboring enemy: indeed, he emphasized a Pan-African as well as a national identity. In our guilt-ridden enthusiasm for multiculturalism we may have forgotten that the rights of minorities rest on systems that depend upon the prior forging of an overriding sense of common nationality.

In a very few societies the political process of ethnic polarization may have gone so far that separation into independent states is indeed the only answer. However, it is a path that could easily lead to the proliferation of tiny states. Consider the latest candidate for statehood, Kosovo, which is a landlocked, resource-scarce, tiny, war-scarred territory. Three tiny territories in the vicinity of Kosovo are also claiming statehood and would presumably use it as a precedent: Abkhazia, population 200,000; South Ossetia, population 70,000 and landlocked; and Transdniestra, population 550,000 and landlocked. Globally, at the last count there were seventy such claims. Most of them make Yorkshire look huge.

If nation building is not feasible, then perhaps Canada and Belgium offer an alternative. These are both strong states in societies in which the sense of national identity is weak relative to the sense of subgroup identity. There is so little common national feeling that both of these societies periodically teeter upon the brink of breaking apart as states. Yet both countries function brilliantly: Canada is at the top of the Human Development Index and Belgium is among the richest countries in Europe. Their intense subnational identities are made manageable within a single state by robust accountability: checks and balances keep the federal state impartial despite the intergroup contest. Instead of a shared sense of belonging, the state functions because its component groups are suspicious of each other and can use the institutions of accountability to prevent being disadvantaged. Such societies may not be cozy, but they are viable.

But here is the problem: Canada and Belgium work because they each have robust systems of accountability. How did they acquire accountability despite the problems that are usually encountered in generating public goods in divided societies? Given their locations, cultural affinities, and size relative to their neighbors, I think that the most likely explanation is that they adopted the neighborhood norm of accountability. In effect, they were free-riding on the norms developed in neighboring societies that had forged a stronger sense of nationhood. The societies of the bottom billion are not in neighborhoods that have the norm of accountability. Given their neighborhoods and their internal divisions, they have not been able to generate the robust systems of accountability that would be needed for them to function like Canada and Belgium. The sequence of introducing elections before either accountability or nation building has been fundamentally flawed. In the now-mature democracies the sequence was reversed: critically, accountability was in place well in advance of competitive elections.

In the absence of accountability electoral competition actually impedes its subsequent supply. The society becomes more polarized and incumbents use strategies of power retention that require them to keep accountability at bay. Unless the states of the bottom billion can forge themselves into nations they will need some deus ex machina that introduces accountability. But where might such a deus ex machina be found?

Chapter 9

BETTER DEAD THAN FED?

I T IS TIME FOR THAT deus ex machina. The key idea is that a minimal international intervention could unleash the powerful force of the political violence internal to the bottom billion as a force for good instead of harm. As such it recognizes the reality that the scope for robust international action is very, very limited.

Even minimalist international intervention needs justification, and so I start with the case for the international supply of key public goods. I will focus on the two that are surely the most important: accountability and security. They are, however, by no means the only public goods that will need to be supplied internationally. Accountability and security are vital: without them a country cannot develop. The societies of the bottom billion have not, individually, been able to supply either accountability or security. The path of building supply from within the society is hard. While its heroes who are engaged in this struggle deserve our support, we should be far more forthcoming with international supply. I will argue that a minimal degree of international intervention could spring the trap. Once the trap is sprung, domestic supply could and should replace it: international assistance in the supply of accountability and security need only be a phase.

There are two distinct reasons that these public goods should be supplied for the societies of the bottom billion internationally, rather than by their own national government. One is that such internal supply has not proved feasible: as you have seen, these societies are usually too fragmented to achieve the necessary collective action. Now I want to introduce a further reason. Because the typical country is so small, many of the externalities that are the basis for public goods cannot be internalized at that level because they spill over to the neighborhood. Indeed, since the pertinent scale of a country for the supply of public goods is its economy rather than its population, the typical country of the bottom billion is far smaller than it might appear. The national income of Luxembourg, the joke tiny country of Europe, is around four times that of the average country of the bottom billion. Public goods that are national in most other societies are regional across the bottom billion. What can be supplied nationally in India would need to be supplied regionally among the plethora of states that make up West Africa or Central Asia.

The most critical missed scale economy due to small size is security. In the countries that are now high-income, the Darwinian process of state selection through violent contest produced countries that were large enough to supply security. Following their economic growth, most of these countries are now also large enough to supply a wide range of public goods at the level of the nation-state. In contrast, the countries of the bottom billion are mostly *too small to be states*. The problem of being too small is, if anything, even more daunting than the problem of being too large. If a continent is divided into a patchwork of tiny countries each too small to have internalized the key externalities, vital public goods will be missing. Even for basics, such as the generation of electricity and the provision of road and rail networks, in a patchwork of small territories the public goods are regional rather than merely national. The radically larger scale of territory of the colonial empires is one reason that their infrastructure decisions were manifestly superior to those of the post-independence governments: Africa is still relying on their faded legacy.

To get specific, Central Africa has ideal geography for hydro-electric power: high rainfall over a massive area of high ground that collects into the River Congo. The descent to sea level could generate power for much of Africa and has been a development project for decades. The project has barely moved. The Democratic Republic of the Congo does not itself need all that power, while other countries are not willing to put themselves at the mercy of its president, or, for that matter, at the mercy of the presidents of any of the countries that power lines might have to traverse. The excess of national sovereignty possessed by these presidents has delivered power shortages across the region. As a huge landmass Africa is also well suited to railways. Many were built by the colonizing powers. Try traveling on them now: there is an acute shortage of rolling stock. It should be easy to finance new rolling stock: elsewhere a rail company can raise finance by pledging the stock itself as collateral, much as you can buy an auto on credit. But the rolling stock cannot be accepted as collateral because it might roll away over a national frontier. There is so little neighborhood cooperation in law enforcement that once it is across the border it might as well have been taken to Mars.

So small is ugly as far as public goods are concerned. Being small artificially limits the benefits of state provision, and this accentuates the lack of supply: the lower the payoff, the weaker the incentive to try.

POTENTIALLY THE STATES OF THE bottom billion could themselves cooperate to supply the public goods that cannot be supplied at the level of each state. Indeed, to the extent that they cannot efficiently be supplied because they are regional public goods, there is an incentive to cooperate. Regional cooperation is the least invasive challenge to national sovereignty, and so if it is feasible, it is at this level that international supply of accountability and security should be undertaken. Is it feasible?

Since the societies of the bottom billion are radically less able to supply the key public goods at the level of each state than are other societies, it might be expected that they would rely more than other societies upon cooperation. They have much more to gain than the larger and more homogenous high-income countries. Among the high-income countries themselves this is indeed the clear pattern: the two countries that have been least interested in pooling sovereignty are the two largest, America and Japan. The country that has been most enthusiastic, indeed providing the home for the European Community, has been the small and diverse society of Belgium. I almost forgot Luxembourg, which is equally keen. Among the plethora of countries that emerged from the collapse of the Soviet Union, the small countries have queued up to pool their sovereignty into the European Union, whereas Russia has held itself aloof.

Accepting these understandable differences in the degree of enthusiasm, over the past half century the developed nations have started to get the hang of how to cooperate, albeit fitfully. Gradually, sovereignty is being pooled where there are clear advantages. The most dramatic pooling of sovereignty occurred earlier: the shift toward federal power *within the United States*. Fifty states, nearly all of which have economies far larger than the typical economy of the bottom billion, have learned how to cooperate. The next most dramatic is the European Community: twenty-seven states have pooled some sovereignty, although much less than in the United States. At another layer down, the Organization for Economic Cooperation and Development is a grouping of thirty high-income countries that has built up a long tradition of mutual reinforcement of governance.

Even the middle-income countries have no equivalent to the sovereignty pooling of the high-income countries. The Asian tsunami was so devastating because the countries bordering the Indian Ocean had not got around to cooperating on an earthquake-warning system. In the bottom billion the lack of cooperation is more pronounced. There are many regional groupings of these countries but they do not effectively bind their members: they are essentially decorative.

To see the contrasting trends, compare Germany, the largest country in Europe, with Burundi, one of the smallest countries in Africa. Both countries have a troubled past and have been a menace to the neighborhood, but think how their degree of sovereignty now differs. One of them does not have its own currency, does not control its own interest rate, does not control its own trade policy, is subject to rules that limit its budget deficit, can have decisions in its courts overruled by decisions in courts run by the neighborhood, and cannot prevent foreign companies from taking over its firms. The other country has total sovereignty over all these matters. The country with the more limited sovereignty is Germany: yet the German economy is thirty-two hundred times the size of the Burundi economy. If we apply the concept of internalizing externalities, Burundi should be pooling its sovereignty with its neighbors far more vigorously than Germany. Generally, small countries need to pool more sovereignty than large countries. Everyone other than Americans gets upset that America often refuses to pool its sovereignty, but as the largest economy in the world America has least need to do so: it has already internalized a huge array of externalities by pooling sovereignty within its borders. The paradox is that despite having the most to gain from pooling their sovereignty, the societies of the bottom billion have pooled it the least.

Return, for a moment, to those externalities that each country of the bottom billion has on others. Sometimes these externalities are reciprocal, so that if two countries cooperate they both benefit. These are the easy public goods to supply through regional cooperation. If everyone benefits, then cooperation should be feasible, although even here the record is not encouraging. But often the externalities are not reciprocal. Quite commonly, in the absence of cooperation, although one country suffers a lot from adverse externalities, the other country gains a little. If I were to write through the night to music, it would help me a little but my family would not be able to sleep. In a family it is easy to internalize that externality: I write in silence. But if Kenya were to fix the road to Uganda and commit to keeping it open, something that would help landlocked Uganda enormously, it would cost Kenya a little money and Kenya would have to sacrifice some political leverage. These public goods are unlikely to be supplied by cooperation. In principle, economics has the solution to such situations: the government of Uganda should offer sufficient financial compensation to the government of Kenya that cooperation is in the interests of both countries. It doesn't happen and it is not going to happen: the road linking Uganda to the coast has been unreliable ever since Kenya's independence. Or take the new iron ore discovery in Guinea. Fortuitously, an existing colonial railway links the site to a nearby port, Buchanan. But unfortunately, Buchanan is in Liberia, so instead a new railway is to be constructed and a new port built. The new route will be much longer, but it will stay within Guinea. More than half of the $6 billion cost of the new mine is due to this decision: the extra cost about equals the national income of Liberia. For those externalities that are not reciprocal, regional cooperation is ruled out, and so the only option is to internalize them at a higher-level international cooperation.

This is by no means the only problem faced by regional cooperation. Consider specifically the provision of the missing public good of accountability. African states are indeed currently cooperating to provide a degree of mutual scrutiny through the African Peer Review Mechanism. This is a new arrangement whereby governments can volunteer to be assessed by other governments. I strongly support it, but to date African governments have shown no stomach for such criticism. Indeed, such an approach faces enormous difficulties. If, within their own societies, none of the component governments is individually accountable, a club to provide accountability regionally faces two acute problems: legitimacy and incentives. The legitimacy problem is that the first government to be criticized can turn around and say, quite reasonably, "So the pot is calling the kettle black!"

The incentive problem is that interstate cooperation largely means inter*government* cooperation. But why should governments that are not accountable cooperate to build restraints upon themselves? Even if some governments are sufficiently farsighted to see some gains from such restraints, cooperation can usually be blocked by a few stubborn participants. Take the recent collapse of accountability in Zimbabwe. If ever there is going to be a need for accountability to be reinforced by the neighborhood, this is it. To his credit, in 2007 President Mwanawasa of neighboring Zambia indeed tried to raise concerns about the meltdown in Zimbabwe at a meeting of the presidents of southern Africa. With several million Zimbabweans fleeing the country his concern was understandable. But President Mwanawasa received little support from other presidents. Indeed, the report comparing economic performance that had been prepared for the meeting was not even presented, lest it cause embarrassment. Mugabe himself stormed out of the meeting as though the very expression of concern was an outrage: why should he not ruin his country if he wanted to do so? Indeed, African presidents have generally rallied around President Mugabe. Far from criticizing him they elected Zimbabwe to the chair of the United Nations Human Rights Committee. Even when Mugabe tried to import a huge arms shipment, it took a strike by South African dockers to block it. Mugabe could only have wanted the guns either to crush the opposition or to menace his neighbors, yet without those dockers, neighboring governments would have remained passive.

A final problem is that just as leadership matters in forging a nation out of its distinct ethnic groups, so leadership matters in galvanizing a group of countries into meaningful collective action. The European Union did not just happen: it took the vision of committed leaders who saw that the long-term interests of their country would be enhanced by pooling some sovereignty. So it is the responsibility of the national leaders within the political groupings of the countries of the bottom billion to build cooperation. In recent years they have been rather short of visionary and charismatic regional leadership. The last time Africa had such leadership was in the early post-independence phase when Presidents Nkrumah of Ghana and Nyerere of Tanzania promoted an agenda of Pan-Africanism. The actual content was a prisoner of its time: the agenda of unity against the Western world. But Pan-Africanism failed not primarily because of this content, but because forging any unity of purpose among so many countries is difficult.

IF SOVEREIGN STATES ARE TOO small, yet regional cooperation between them is too difficult, one radical alternative is to federate them into a few larger states. This is the route that America took, and it was briefly tried in late colonial Africa. A straightforward obstacle to merging states is that certain personnel would become redundant, and like all about-to-be-made-redundant employees, they might object. If two states merge they need only one president, one set of ministers, one army. Perhaps that is why state mergers are so rare.

While the perks of high office might account for the reluctance of states to merge, there is a deeper question: would it alleviate or intensify the key weakness of small size, the inability to reap the scale economies of security. However, recall that unfortunately there are two opposing effects of an increase in size on insecurity, not just the benefits of greater scale but the increased risk associated with any consequential reduction in cohesion. But would state merger further worsen the lack of cohesion? It is quite possible that given the arbitrary nature of colonial boundaries, with straight lines carving through societies, some mergers would not increase ethnic diversity. Even if it did, the security gains from scale might outweigh the dangers of increased diversity.

This is a question that is just about researchable. Christian Wig-strom, a Swedish graduate student at Oxford, got interested, and we decided to investigate it. We decided to replay the decolonization process, imagining the consequences had Africa been packaged into fewer countries. Rather than redraw boundaries, we decided to pro-gressively rub them off the map, first merging countries into pairs, and continuing the merger process until we reached the African dream of a politically united sub-Saharan Africa. You may variously view this as social science gone mad; as an arrogant attempt to play God with countries; or as something that might conceivably inform the political impetus within Africa to break beyond the colonially imposed mosaic and achieve greater unity.

One of the most exhilarating consequences of building a model is that it enables the researcher to simulate alternative scenarios. We had to establish some principle to guide the merger sequence: for example, should Kenya first merge with Uganda or with Tanzania? We decided that our guiding criterion would be to minimize the risks facing the merged state, so we looked for states with similar ethnic composition. In effect, these were hypothetical marriages of countries that were as ethnically similar as possible. In the process we discovered that because the boundaries of the old empires often sliced through ethnic groups, it was sometimes possible to make na-tions bigger without making them more diverse. On our analysis such nations would have been more secure: they would have gained from scale economies without losing from additional diversity. We also found that state boundaries at least appear to have been drawn as attempts to trade off scale against diversity. In places where there was atypically high ethnic diversity, the states were also unusually small. It is these states, small yet diverse, that face the most severe problems of internal security. As the hypothetically merged nations gradually came out of this process, we then estimated their risks of violence from the simulation model that I had built with Anke and Dominic. We could address, at least after a fashion, the question of how badly wrong the decolonization process had been in their packaging decisions from the perspective of security.

Our somewhat eccentric research is still in progress, but it looks as though Africa's multitude of ethnic groups could have been bundled up into around seven large states with little increase in ethnic diversity. A seven-state structure for Africa would, on our analysis, have been considerably safer than the present structure. However, to crunch down from seven states to one—a United States of Africa—would involve a high price in terms of increased diversity and so drive the region back into danger. Perhaps the goal of greater African unity may be best achievable through strengthening the subregional groupings.

GIVEN THE SEVERE PROBLEMS OF cooperation among the sovereign states of the bottom billion and the impediments to the more radical strategy of state merger, the only remaining option is for provision to come from a higher level of international cooperation than the region. Those societies that are currently being damaged by rule-free electoral competition have an urgent need for accountability, and it will need to come from others. More specifically, because of the problems of legitimacy and incentives, it will need to come predominantly from that part of the international community in which governments are already subject to effective accountability. We are back to the brick wall of national sovereignty, reinforced by the mindset that resulted in the election of Zimbabwe as chair of that human rights committee.

Conventionally, the governments of the bottom billion are regarded as internationally powerless. They see themselves as victims of an international system that is stacked against them. Having struggled free from being colonies they see themselves as still entrapped by the bullying of more powerful nations. I think that this victim-bully imagery has been hugely dysfunctional. It has masked a radically different reality: individually, the governments of the bottom billion have too much sovereignty, not too little. Before the people I most want to reach throw *Wars, Guns, and Votes* down in disgust, let me stress that I am not an apologist for colonialism, and I most certainly do not want to restore it in any shape or form. The problem I want to address is first and foremost a problem *for the societies of the bottom billion themselves*.

The countries of the bottom billion are, for the most part, the opposite of America. Rapidly put together in a surge of immigration, America was an instant society but is now an old nation. Americans share not only a sense of identity but a suspicion of governmental power and so have cooperated to build and sustain the public good of checks and balances: government is highly transparent. America has also been expansionist and so is now enormous, bringing with it the scale economies of security. The societies of the bottom billion are ancient, but as states they are instant. Their states are usually too small to reap adequate scale economies of security, so they struggle to keep the peace within their societies. Because they are instant, they have seldom forged strong national identities to compete with their ancient social identifiers of ethnicity and religion. As a result, although too small for security, they are too large for the social cohesion that is hugely helpful for the provision of public goods. So public goods are in short supply.

As we have seen, one missing public good is the accountability of government: in contrast to America, the governments of the bottom billion are not subject to many internal checks and balances. If the societies of the bottom billion cannot supply themselves with this public good, then it is better supplied internationally than not at all. The argument is analytically equivalent to the provision of a vaccine against malaria, another enormously valuable public good that is missing. No society of the bottom billion is able to surmount the difficulties of providing this public good, and so we rightly look to international action to fill the gap. *The public goods that benefit a region are sometimes best supplied outside the region.* The difference between supplying the missing checks and balances and supplying a vaccine against malaria is national sovereignty. A vaccine against malaria developed through international public finance does not challenge national sovereignty; checks and balances developed through international public action do.

The most enduring legacy of the colonial experience is the excessive respect given both within the societies of the bottom billion, and by those who are concerned about their fate, to the notion of national sovereignty. The sentiment "never again" impedes serious thought. In reality, the typical society of the bottom billion does not have *national* sovereignty. It has yet to become a nation as opposed to a state: so it lacks the cohesion needed to produce effective restraints upon either the conduct of elections or the subsequent power of the winner. As a result, it has *presidential* sovereignty. No wonder presidents are jealous of national sovereignty: they are jealous of their own power. The key struggles, that for accountability, that for security, and that for better provision of the more conventional public goods, all depend upon facing down the shibboleth of national sovereignty by recognizing it for what it really is. There is no shame in meeting these needs internationally: they are far better met internationally than not met at all.

The international provision of accountability to the rule of law faces a standard objection: fairness. Why should some societies subject themselves to international rules if others won't? To be specific, if America won't subject itself to international rules, why should East Timor? This sentiment is understandable, but it is fundamentally wrong: it is part of the mentality that blocks serious thought, so let's address it head on.

I would indeed like to see America more supportive of international rules: there are some global public goods from which it would benefit and that even America cannot supply by itself. But quite clearly America's citizens have radically less need to subject themselves to international rules: as a nation it is already supplied with restraints on government; as a large state it can already supply its own security and a huge range of other public goods. In contrast, the citizens of East Timor need to rely on international rules because they are living in a territory that is structurally unable to meet these needs at the level of the state: at present thousands of them are cowering in refugee camps for fear of one another. The citizens of East Timor would, like Americans, benefit from the global public goods, but they can potentially gain far more than this from international rules. The purpose of sovereignty is not to be a virility symbol with which presidents strut on the world stage, it is part of the design of government: the criterion should be the needs of citizens. The elite passion for sovereignty at the expense of need amounts to "better dead than fed." It sounds quite noble until you realize that it is not the elite who go hungry. Today, as I make my final revisions to the manuscript, the phrase has literally come true: President Mugabe has banned food aid to his starving country. Errant voters will be starved into support.

Quite probably, East Timor's need for strong international rules is only temporary. Once such rules had successfully supplied its citizens with the accountability and security that they now lack, the society and economy would progress. As it did so, the tide would start to flow in its favor. In ethnically diverse societies, as long as the easy options for winning elections are closed off by enforced rules, democracy does deliver faster growth. Recall that once growth takes a society above $2,700 per capita, democracy also begins to make it more secure. With time, checks and balances that are initially internationally enforced can become internally sustained. Just as one coup legitimizes the next, so the accumulated history of adherence to rules builds the practice of compliance.

The decolonizing goal of a world of nations, sovereign and equal, was surely right. It is preferable to supply as many public goods as possible nationally, rather than internationally. It is known as the principle of subsidiarity: sovereignty is best lodged at the lowest level needed to achieve its function. States with small populations eventually become viable once they have reached a high level of income: there economies can be quite big, and they learn to integrate with neighboring countries so that a relatively small size does not inflict significant costs. Luxembourg is the richest country in Europe and can at some point provide a model for even the tiniest countries among the bottom billion. But the abrupt transfer of sovereignty from the near-global remittance of the empires to the presidents of states that were tiny yet diverse, while reflecting the right goal, took the wrong route. It condemned many little countries to catastrophic underprovision of the two vital public goods: accountability and security. A phase of international provision of these goods is needed for these societies to reach the goal.

I am now going to focus on the need for accountability. To break the impasse a phase of international supply is needed, but is it realistic? The international community is about as dysfunctional as a community can get. Its core of jelly is no match for the ruthlessness of an incumbent politician. The key move in this book is to harness the one force that genuinely has the power to discipline them. There are two critical dimensions in which a government needs to be held to account: rules determining how power can be acquired and rules determining how power, once acquired, can be used to spend public money. How, in practice, could the international community introduce effective rules?

PROPOSAL 1: HARNESSING VIOLENCE FOR DEMOCRACY

The legitimate route to power is through an election that is free and fair. As the Kenyan elections of 2007 demonstrated, the societies of the bottom billion are not themselves able to supply the vital public good that restrains electoral malpractice. Kenya has long been regarded as the most advanced country of Africa:

if Kenya cannot do it, few can. And so it must be supplied as an international public good. The problem has always been how.

Quite clearly, the international community cannot enforce democratic standards on a country whose government is unwilling to adopt them. This is the jelly problem and it is not going to change. At present it poses a debilitating dilemma for donor governments. However badly the government of Kenya behaves following elections, at least it has held an election. It is manifestly better than many other governments of the bottom billion. Donors feel that they can hardly suspend aid programs to Kenya while they continue providing aid to all these other countries. Understandably, they feel that they cannot impose double standards, judging Kenya by tougher criteria than the rest. Yet double standards are precisely what are going to be necessary: here is how to do it.

Proposal Version 1:

This is a proposal that might be drawn up by any concerned idealist. It is for a voluntary international standard for the conduct of elections, linked to a powerful carrot. Governments could then choose whether to sign up to the standard, entitling them to the carrot. Once a government had signed up, it could be monitored, rewarded, and punished on a different scale from the rest.

Reader, I can hear you saying, "What a great idea; let's move on to the next problem. Oh, incidentally, what would be the carrot?"

Proposal Version 2:

The core of an effective proposal is to design that carrot. To be effective it would need to be big, but above all it would need to be credible. When I explained my idea to a wise old practitioner, he interrupted at this point: "Don't suggest aid as the carrot: after decades of donors not sticking to their conditions, it just isn't credible." Quite right: the carrot is not going to be aid, it is security. The international community is going to provide a guidance system that transforms the missile of the coup d'état into an effective domestic restraint on misgovernance.

Key members of the international community would make a common commitment that should a government that has committed itself to international standards of elections be ousted by a coup d'état, they would ensure that the government was reinstated, by military intervention if necessary. This carrot is in itself not negligible: remember that presidents face risks from a coup that are far higher than those they face from an election. And remember that democracy alone does not strengthen the defenses against it. There have been more than eighty successful coups just in Africa, versus a mere handful of electoral defeats. But, as you will see, the key aspect of this carrot is that it turns into an equally powerful stick. This carrot-cum-stick may be sufficiently big, but is it sufficiently credible?

At this point modern economics becomes surprisingly useful in helping us to think through whether standards linked to coup protection would be effective. The method it uses is a game tree. There need be nothing fancy about thinking through a game tree: essentially it is a matter of repeatedly posing the question, "And so if I did that, what would you do next?" The insight brought by economics is that although you first have to set down the game as a sequence of "What would happen next?" you *solve* it by reversing the sequence, starting with the last decision that has to be taken.

So first let me sketch the game of voluntary standards for democratic elections. The decision tree has many branches, but for standards to work, one particular branch is critical, and that is the one on which I will focus.

Step 1:

The international community promulgates a voluntary standard for the conduct of elections. It is entirely voluntary, but governments that feel the need for enhancing their democratic credibility can choose to commit to it. If the government commits itself, it is rewarded with a counterpart commitment from the international community. The commitment is to put down any coup against the government, by military force if necessary.

Step 2:

The government of a bottom billion society now decides whether to sign up. If it decides not to do so, end of story.

Step 3:

If the government commits itself, then there are various possibilities. The important one is what happens if it subsequently finds that it is heading for defeat in an election. At this point the government must take the decision whether to abide by its commitment to the international standard, or breach its commitment and steal the election.

Step 4:

If the government decides to break its commitment by stealing the election, then the ball goes back into the court of the international community. It must decide how to respond. It can, if it chooses, publicly declare that the government has breached the standards for conducting a democratic election and withdraw the commitment to put down a coup.

Step 5:

If the international community withdraws its commitment to put down a coup, then the ball flies out of the court altogether, and lands at the feet of a new player: the military. The military has to decide whether to launch a coup.

Step 6:

If the military launches a coup, the ball goes back to the international community. They can turn a blind eye and just ignore it; they can condemn it; or finally they can welcome it on condition that the leaders commit themselves to hold internationally verified elections within a specified period.

Step 7:

The ball finally flies back to the coup leaders. If the international community welcomes the coup subject to conditions, they must decide whether to accept these conditions and proceed to free and fair elections, or to cling to their new power: whether to be Colonel Vall of Mauritania, or General Gueï of Cote d'Ivoire.

Now we solve the game by working through it backward. Start with step 7: will the coup leaders abide by the conditions set by the international community? If they do they are heroes and can be treated accordingly, if not, as you will see, they are living dangerously. The coup leaders will have come to power precisely in the context of a stolen election, and this is surely the rationale that they will have used to motivate their own soldiers into action. Coup leaders are not inevitably self-serving like General Gueï. Mauritanian coup leader Colonel Vall promptly arranged free and fair elections and stood down. But suppose that, having said all these things during the coup, the new leaders then became so partial to power that they refused to hold verified elections. What might happen then? The answer is that they themselves would face a high risk of a further coup. Remember the risk is high because one coup leads to another. The leaders of a second coup would have a ready-made justification, and the leaders of the first coup would face dire consequences: they would have no protectors. Gueï himself suffered a countercoup once it was clear that he had reneged. So here is the punch line: *the threat of a further coup enforces the commitment to the intervention being temporary*. Of course, people sometimes make mistakes or they take big gambles that pay off. But in step 7 the most probable outcome is for the coup leaders to abide by the conditions of the international community.

Now we can roll the game up, turning to step 6. Having sorted out step 7, step 6 is pretty obvious. Why would the donors ignore the coup or condemn it, when by responding with conditional acceptance they can pave the way for verified elections?

Now we are ready for step 5. The decision at this step is taken by the military: should it launch a coup? The circumstances, remember, are that the international community has certified that the government has stolen the election and publicly withdrawn its commitment to intervene to put down a coup. The answer is that we really cannot tell whether the military will launch a coup in these circumstances. Perhaps the president has established such an intrusive form of repression that even discussion would be too dangerous. Perhaps the military is entirely dominated by the president's close family and they all love him to bits. But quite possibly the bored general staff decides that this is their moment. Above all, they will worry that if they don't take this decision, other more junior officers will launch a coup instead. In that eventuality the current leadership will be plunged into an ignominious premature retirement. So: *the threat of a rival coup makes a coup more likely*. A close parallel to this stage was the coup threat that enforced regime change following the Senegalese elections of 2000. Remember that the threat was emboldened by the coup in Cote d'Ivoire, which had revealed that the French security guarantee had been withdrawn across La Francophonie.

Now for step 4, which is in fact the crucial step: will the international community have the backbone to stick by its commitment despite its deserved reputation for being pure jelly? Steps 5 through 7 have provided us with the answer. The international community gets what it most wants by the strategy of declaring the elections illegitimate and withdrawing the commitment to suppress a coup. Indeed, it is actually much stronger than that. Suppose that the international community does *not* withdraw the commitment. Now what would happen if the military launches a coup on the argument that the government is illegitimate? The international community finds itself in the dangerous, damaging, and embarrassing situation of either breaking its commitment to put down the coup, or intervening militarily to defend a government against domestic forces for decent governance. No doubt about this one: the international community withdraws the commitment to put down a coup.

At last we are at the step that cuts the mustard: step 3. The government realizes it may lose a fair election: should it steal it? We know what the answer to that is if the government has not signed up to an international standard: look at Nigeria, look at Kenya. Are things different if it has committed itself? We now have the answer. The government thinks through how events would unfold: if you doubt this, remember that the representatives of the international community will be explaining in graphic terms how they would react. More crucially, the presidents of the countries of the bottom billion have all been selected through a Darwinian struggle in shrewdness: they may not always be the sort of people your mother would invite for tea, but they would beat you at poker. They think it through: stealing the election no longer looks such a good idea. This is, indeed, precisely the calculation that confronted President Abdou Diouf of Senegal: step down with honor and dignity, or face a high risk of be ousted by a coup. He stepped down with honor.

And so to the potential killer: step 2. Anticipating all that follows, would any government sign up? We have seen the downside: the government loses the scope to steal an election and this is costly.

So governments will only sign up if there are gains that more than offset these losses. We know the promised gain: protection against coups, but *is it credible?*

To find out, we have to investigate the related game of whether the international community would make and then honor its commitment. Fortunately, this game is not so complicated: technically it is termed a "sub-game."

Step 1:

The international community decides whether to commit to put down a coup in return for a commitment to abide by democratic rules.

Step 2:

There is a coup in a country that has committed, so does the international community intervene and keep its promise?

Remember: work backward, so step 2 is first. Why should the international community keep its commitment? After all, this community can hardly say, "My word is my bond," at least not with a straight face. The answer depends as usual upon the costs and the benefits. There is no doubt that keeping the commitment has serious costs. This is the situation in which "our boys," and indeed one day my own boy, may get sent to somewhere most voters have barely heard of, to put down a coup. As Daniel's father, I do not relish such a prospect. But there are also benefits, and the benefits are potentially enormous. We have found a way of making democracy work in environments where otherwise it deepens many of the problems. Suppose that by the time there is a coup that needs to be put down, a dozen governments of the bottom billion have already signed up to democratic standards.

And now put yourself in the shoes of the politician who has to decide whether to keep the commitment or renege. Am I really

going to be a politician who reneges, wrecking not just this particular country but tearing up the commitment technology that a
dozen countries are already using? If I do that, not only will I be
vilified, I will have to look at myself each morning knowing what
I did. But, despite my posturing before the voters, I am 99 percent
jelly. I decide to ask my military: can we put down this coup? So
what will the military say? Realize that they will have been training
and equipping and extracting extra budget for just this eventuality.
They have already checked up on the historical record of how the
French and the British military put down coups in little countries
swiftly and with virtually no losses: this is not another Iraq. The
chief of the general staff stares down at the politician: "Piece of cake,
sir!" he replies. The ball is back with the politician: at some point
even jelly solidifies. In fact, the game is more satisfactory than I have
presented. Knowing that the coup is very likely to be put down, only
the drunken potential coup leaders make the attempt in the first
place, and so coups become rare and incompetent.

So much for step 2, how about step 1? If it is worthwhile putting a coup down when it occurs, it is worthwhile making the commitment to put it down. The international community gets many
of the benefits immediately, and any costs are in the future. So we
arrive at the solution that the coup commitment is credible and so
the benefits for signing up to the democracy compact are considerable. There is one further reason that the international community
should regard a guarantee against coups as appropriate: think back
to what drives the risk of a coup. Aid significantly increases coup
risk, and so donor governments are inadvertently exposing recipi-
ent governments to a menace. It is a menace they could and should
see off.

It is time to return to the question of whether in these circumstances any political leaders in the societies of the bottom billion would sign up to international democratic standards. Not only
would protection from the domestic military be attractive but the

government may also value other benefits. It gains legitimacy in the eyes of donors, which may translate into cash. It may gain legitimacy in the eyes of its own citizens, which may translate into greater power to achieve its objectives. And finally, there is likely to be a push factor. The political opposition will almost certainly try to gain electoral advantage by publicly and vociferously undertaking to sign up to international standards if elected. The opposition will make this commitment as part of its critique of government unfairness, and because donor support may reduce the extent of government cheating.

Indeed, I think it likely that President Kibaki of Kenya would have committed himself to international standards, had there been any, on first coming to power in 2002. His prior campaign had been based around a series of promises to change Kenyan politics: an international standard would have suited him well. Similarly, I think it very likely that Raila Odinga, the Kenyan opposition leader, would have committed himself to them during the campaign of 2007. After all, following the declaration of the doubtful results, he called for precisely the international intervention that a commitment would have triggered. If the opposition is gaining political traction by making such promises, the government may decide that is it best to neutralize it by making the commitment itself.

Were there to be an international standard, the leaders of the bottom billion would begin to sort into sheep and goats. And that itself would gradually increase the pressure on the goats. But given all these benefits, would any nation or group of nations with the military capability be willing to provide a security guarantee against coups? Only a few nations have the required military logistics for rapid deployment of sufficient force: America, France, and Britain. Are they willing? Well, they already have the forces in place. America is in the process of creating a dedicated military rapid-reaction force stationed in Africa, rationalizing its existing capabilities around the region. Appropriately, while the commander of the force will be a general, the number two will be a development professional. France still has a chain of military bases in West and Central Africa, and Britain is already providing a security guarantee to Sierra Leone. As I write, America is searching for African governments willing to host the base: South Africa and Nigeria, two obvious locations, have both declined.

After Iraq, many governments are naturally apprehensive about the American preemptive use of force and so are wary. South Africa and Nigeria are probably also concerned that an American capability in the region might dilute their own regional superpower status. But the brute fact remains that neither South Africa nor Nigeria is able or willing to provide the necessary military capability itself. Nor, if either developed such a capacity, would it be welcomed by neighbors who are probably more apprehensive of the big brother next door than of the global superpower. Whereas with America it is its recent military behavior that raises concerns, with France and Britain it is the colonial record. The world is not ideal: there is no military power that is untainted in African eyes. But precisely because of these concerns, it is surely better to have these forces bound by clear rules of use. While the governments of South Africa and Nigeria might well not wish to host foreign forces with an unclear mandate, they should welcome them for the specified purpose of protection from coups against governments that have committed themselves to proper standards of democratic elections. "Keep out of Africa" is irresponsible if it condemns the continent to unaccountable government.

Finally, I turn to my most demanding readers: those presidents who, having read the section that sets out strategies for reducing the risk of a coup, still could not sleep soundly. Gentlemen, I promised you that if you read on you would find a fully reliable protection from your own army. You now have it: you no longer have to trust your brother-in-law. All you need to do is to lobby at that otherwise useless international jamboree to which you have just been invited, for a compact on democracy. You make a note of it and fall asleep.

PROPOSAL 2: ENFORCING PROBITY IN PUBLIC SPENDING

Proposal 1 provides some rules for how a government acquires power. Proposal 2 shows how the international community could also feasibly provide some rules for the use of power. At the heart of the abuse of power is money.

Public revenue, whether from aid or from taxation, is not a trough for political patronage: it is there to finance the public goods needed for a society to be decent and prosperous. But, as scandal upon scandal demonstrates, public money is only put to these proper uses if politicians and senior civil servants are shielded from temptation by systems of scrutiny and punishment. In the developed societies in which corruption is now a rarity we tend to forget that the habit of honesty is built on the bedrock of fear of detection. In most of the societies of the bottom billion, systems of public scrutiny were dismantled from the top. The resulting grand corruption not only wasted public resources, it empowered the political crooks. Patronage financed by embezzlement has been the standard means of retaining power. How can international action help to put restraints back in place, given that people who are politically powerful would stand to lose?

For most countries of the bottom billion the answer is quite straightforward: much of the money comes from aid. Donors have both the power and the obligation to ensure that this money is well spent. For many years donors hid behind the illusion that their money was financing specific projects to which it was ostensibly tied. As aid-receiving governments have increasingly been encouraged to determine the content of aid programs, this has become even more of a fiction. Quite evidently, if donors finance the projects that governments want, the chances are that many of these projects would otherwise have been financed out of taxation. There is nothing intrinsically wrong about such a process, but what the aid is actually financing is whatever the government would otherwise not have done. Knowing that the Swedish government is willing to finance schools, the government of Ethiopia, which also wants to expand schooling, requests that the Swedes pay for it. This releases money that the Ethiopian government would otherwise have spent on schools for whatever else it might choose.

As donors woke up to this problem, most of them moved away from projects to budget support: that is, they simply handed over a check to the government, which it could treat as general revenue for the budget. While this acknowledged reality, it was sometimes highly irresponsible. Unless budget systems are sound, money put into the budget will leak into patronage.

It is one thing to say that budget systems should be sound and quite another to ensure it. Two complementary resources are needed: capacity and verification. Public revenue will leak from wherever there is a hole, and so there is a large preliminary task of overhauling the practical processes by which money is spent: budgets need accountants, and lots of them. Starting from a culture in which there is no presumption of honesty, the system of financial checks needs to mirror the paranoia of the dictatorships: there needs to be so much interlocking monitoring that even if a few accountants are prepared to be corrupt, they cannot make a difference. Superficially this may look wasteful, because the administrative cost per dollar of public spending will be much higher than where there is already a prevailing culture of honesty. But that is simply the reality: corruption makes public spending less efficient.

Donors can help governments put proper systems of accountancy into place through technical assistance, which is the phrase used to describe the supply of skilled people. This is the most despised form of aid, but it is often essential. Installing the capacity is not enough: the donors must verify that the capacity is actually being used to enforce probity in public spending. This requires a forensic approach quite distinct from the cooperative nature of capacity building. Only where forensic scrutiny of a budget system has already certified it as satisfactory should aid be provided as budget support. Of course, if the donors introduced that as a rule tomorrow, they would not be able to disburse any money, and so there would need to be due notice and transitional arrangements. But there is simply no alternative: projects are largely an illusion, and so for aid to be effective, it requires sound budgets. If governments want to spend aid money, this should be the condition for its use. I call it governance conditionality, to contrast it with policy conditionality. Donors should not be telling governments what policies to adopt, or, within the range of public goods, how money should be spent. But they have an obligation both to their own taxpayers and to the citizens of the bottom billion not to connive in budgetary processes by which public money is diverted for private ends.

If this were the condition for aid, some governments would decline it, in particular those with large revenues from natural resource exports. The international community has little financial leverage over these governments, and a completely different approach is required to encourage financial probity. That would take us beyond this book. However, there is another category of country where, while the government might be willing to accept it, the task of building an accountable budget is simply unrealistic: the civil service has degraded too far for recovery in a reasonable time scale. Is there an alternative?

Liberia is now ruled by reformers, led by the admirable president Ellen Johnson-Sirleaf. She was preceded by a government so insalubrious that even the normally pusillanimous donors drew the line. When they could no longer stomach what was happening to their money, they threw sovereignty to the winds and introduced a system called GEMAP (Governance and Economic Management Assistance Program), in which the finance minister could not incur expenditure without a counter-signature by the donors. GEMAP is billed as a great success, but it was in reality a despairing reversion to colonialism. I talked to the reforming new minister of finance, Antoinette Sayeh, and she naturally wanted to move on from it. But to what? The donors surely trust Antoinette, but she cannot herself ensure that public money is well spent. A corrupt finance minister can ensure that it is badly spent, but unfortunately the converse does not apply. A minister depends upon the staff of a ministry. The first act of President Johnson-Sirleaf was to dismiss the entire staff of the ministry of finance. She was right to do so, but what do you do the next day?

Instead of the desperate reactive policy of GEMAP, used only when the situation is manifestly out of hand, the international community needs to anticipate that in some situations conventional systems of accountability have degraded beyond rapid recovery. Since these are the situations in which needs are most desperate, an institutional design is required for how big money can nevertheless safely be spent on the provision of basic services.

The model that virtually all the newly independent governments of the bottom billion adopted was, unsurprisingly, the prevailing European model of the 1950s: monopoly supply by government ministries. Even for Europe this model has proved a little problematic and there has been a move away from it, but for the societies of the bottom billion it was usually inappropriate and in some disastrously so.

A more realistic design would separate the functions conflated into these grandiose ministries: overall policy, the allocation of money to specific activities, and the actual supply of activities. The ministry should be responsible only for overall policy. Indeed, only once policy is separated from the spending of money is the ministry likely to give policy serious consideration: at present attention is often driven by the scope for kickbacks. Where needs are dire and public systems have failed, the actual supply of basic services, such as running a school, should be open to all who can do it well. This is likely to mean churches, NGOs, local communities, and most promising of all, the new philanthropists, rather than government-run operations. I am enormously impressed by the professionalism, innovation, and energy shown by the new philanthropic organizations, typically staffed by young people trained in business schools, who combine passion with an eye for cost-effectiveness.

In between the ministry and these suppliers would be an agency that handled the money: contracting with the suppliers and monitoring their performance, but working to objectives set by the ministry. This type of agency is the missing link: at present, governments spend the donor money and each NGO does its own thing, disconnected from public supply and largely unaccountable. The linking agency would enable donors to channel money to effective service delivery. In return, donors would share oversight of the agency with the government and local civil society. Would donor money channeled through such an agency still face the problem I sketched with projects? That is, if donor money funded the priorities of basic social spending, wouldn't this simply release government tax revenue for other purposes? The answer is that in the broken societies like Liberia the economy has collapsed and along with it government revenue. In such societies there is an enormous need for aid, yet at the moment they get very little. They get little because donors know that public systems of spending are too dysfunctional to be used. Donors need to face up to the logical implication: try a new system.

Would the governments of broken states accept such spending systems? If they were accompanied by the realistic prospect of vastly scaled up inflows of aid, I think most would.

PROPOSAL 3: THE INTERNATIONAL SUPPLY OF SECURITY

It is time to turn from accountability to security: this is the public good par excellence for which the states of the bottom billion are too small. Manifestly, its provision has been inadequate: that is why these states are so dangerous. For some public goods the lack of social cohesion that is endemic to the states of the bottom billion can be overcome by decentralization: if at the national level the public purse is viewed merely as a trough, it may be better to forgo some scale economies to gain a sense of common identity. But security is not one of those public goods: decentralized provision of security services would magnify the risks of civil war because it would create rival military forces controlled by rival politicians. Security services need to be supplied not on a smaller scale than the nation, but a larger scale.

As with the other public goods, the high-income countries have been cooperating regionally on the supply of the public good of security for more than half a century. NATO is such a force providing mutual guarantees. Could neighborhoods within the bottom billion do the same? As I will set out, there is plenty of untapped scope for security cooperation among the bottom billion. But before going deeper, is this idea so wild as to be unthinkable? The United Nations has recently formulated a far more radical proposition: *the responsibility to protect*. R2P, as it is known in the preposterous shorthand of international parlance, is a full-frontal assault on the concept of national sovereignty. It proposes that the international community has the right to intervene to protect citizens from their own government. Compared with that proposition, the idea that I am suggesting is decidedly modest. At its most minimal it is that neighboring states would get mutual benefits from binding themselves to security cooperation. At its most ambitious it is that the neighbors of a country at risk of civil war have a right to protect *their own citizens*.

It may be best to start by stating clearly what I do *not* have in mind. I am not proposing that the UN cavalry should ride in to topple President Mugabe, or to impose peace in Darfur. I regard these as distracting fantasies that impede security cooperation for less controversial and therefore more realistic purposes.

So let's begin with the least demanding: security cooperation that generates mutual gains. Think back to neighborhood arms races. If the neighbors are seen as a threat, then the military spending of each government becomes a regional public bad. Arms races do not enhance overall security; they waste money. Worse, as I showed, since some of the guns leak into the informal market, the higher the military spending in a neighborhood, the easier it is for rebel groups to get hold of guns: cheap guns increase the risk of civil war.

Could the arms races in Lilliput be unwound in Africa, as President Arias is trying to do in Central America? An analogy is the mutual deescalation of tariffs negotiated through regional trade agreements. African governments have been negotiating regional trade agreements for years, but there is no equivalent for military spending. One reason is that there are too many players. Suppose that there was an arms race on an island divided into just two countries. It would be straightforward to negotiate de-escalation because the benefits are fully reciprocal: each country's spending threatens the other.

Unfortunately, mainland Africa is at the opposite end of the spectrum from such an island: there is a chain of neighborliness with forty-seven countries sharing the same landmass.

Zimbabwe is next to Zambia, but Zambia is next to the Democratic Republic of the Congo. So the Democratic Republic of the Congo is a potential threat to Zambia but not to Zimbabwe. In turn, the Democratic Republic of the Congo is next to Chad, and so on. With a trade agreement it is possible to cut tariffs against some neighbors but not others: Zimbabwe and Zambia could negotiate with each other and reach a deal that excluded the Democratic Republic of the Congo. But if a country cuts its military spending, this benefits all neighbors regardless of whether they reciprocate. If Zambia cuts its military spending, both Zimbabwe and the Democratic Republic of the Congo benefit. If only Zimbabwe reciprocates, then Zambia has become less secure because it is now weaker relative to the Democratic Republic of the Congo. So for a negotiated reduction in military spending to succeed, everyone has to agree to do it together. Since it is a case of all or none, with so many countries in the region the inevitable result is stalemate. The cooperative provision of security sounds attractive but is extremely difficult.

Given that military spending is at least in part a regional public bad, it should be discouraged. Conceptually, the right way to discourage a public bad is usually to tax it: this is the principle behind carbon taxes. So if only the African Union could reach agreement, it should tax military spending, just as the Eurozone now taxes the regional public bad of excess budget deficits. Of course, the African Union is a long way from being able to initiate such regional cooperation, so is there an alternative? Recall that a public good that benefits a region may not most efficiently be produced *in the region*. Such is the case with the discovery of a malaria vaccine: the benefits would be pan-regional, but the research needs the skill base found only in high-income countries. That is what makes the finance of the research for a malaria vaccine a particularly good use of aid. The same may be the case for the regional public good of reduced military spending induced by a tax on military spending.

By linking aid to the level of military spending, donors could simulate a regional mutual tax. Recall, donors would have a very powerful justification for doing so: to date around 40 percent of military spending has inadvertently been financed by their aid. Donors are subsidizing military spending when they should be taxing it! At present donors often try to discourage military spending by huffing and puffing. It would be both less intrusive and more effective if their justified dislike for military spending was embodied into some clear rules of aid allocation: for example, starting from where military budgets are now, each dollar of increase would be taxed by a 40 percent reduction in aid, which would be redistributed to other countries, and each cut in spending would be correspondingly rewarded. Unlike huffing and puffing, this could not be regarded as an infringement on sovereignty: it provides a regional public good. And being a clear and stable incentive, it would most probably be more effective.

Now let's get more ambitious. So far security provision as an interstate public good has involved mutual gains. All governments want the same public good and security, and international provision improves the technological possibilities of supply. But remember that music: my pleasure might be your pain, or more fancily expressed, my decisions might inflict negative externalities on you. Those externalities need to be internalized even if I don't like it. We now get rather tougher on national sovereignty.

Historically, the entire concept of national sovereignty arose out of the catastrophe of the Thirty Years War, during which governments of rival religious allegiances fought it out to impose their preferences over each other's territories. By the end of the war the astronomic costs of warfare were better appreciated and the bloody game of religious conversion by conquest was recognized as not worth the candle. In its place came the principle of national sovereignty: whatever wrongs a government perpetrated on its own population, they did not have sufficient consequence upon the well-being of other countries to warrant intervention. At the time the concept was developed in the seventeenth century, there were reasonable grounds for such a proposition: economies and societies were not highly integrated. Whether or not it was true at the time, it certainly is not true any longer. Nowadays a civil war generates externalities for neighbors that are too large and too adverse to be dismissed.

I have tried to measure them through studies with Anke, Lisa Chauvet, and Alberto Behar. The approach we used was standard, although care has to be taken to distinguish those neighborhood effects that have nothing to do with war from war itself. For example, a neighborhood might be affected in common by a drought, as in Southern Africa during the mid-1990s. We find, unsurprisingly, that the costs of a civil war to any particular neighboring country are considerably less than the costs to the country itself. Typically, a country might lose around 0.9 percentage points off its growth rate if one of its neighbors is at war. However, the typical civil war country has three or more neighbors, and, further, the economies of the neighboring countries are usually larger than that of the civil war country itself. This is because, as we have seen, being small and being poor are both risk factors.

In our analysis we include only costs to immediate neighbors. This omits demonstrated adverse spillover effects across a wider area. Even with the restriction to immediate neighbors, the numbers imply that the costs to the neighbors as a group are likely to be even larger than the costs to the country at war. So, reflecting the standard economic solution to the problem of how externalities should be internalized into the decision process, decisions that substantially affect the risk of a civil war should be internalized among the neighborhood. In the Democratic Republic of the Congo several of the neighbors got involved: indeed, three of them, Rwanda, Uganda, and Angola, sent their troops into the country. The neighborhood dimension of security could scarcely be more graphically illustrated.

Recall that by far the most dangerous situations are postconflict. Post-conflict relapses are likely and inflict high costs upon neighbors. Should the neighbors of a post-conflict state have some rights to a say in post-conflict policies? That was where my thinking had reached a year ago: post-conflict countries should go through a phase of sharing sovereignty with their neighbors until they had progressed out of danger. And then I woke up to two insuperable problems.

Problem number one: Not only do neighbors have a legitimate interest in the governance of the post-conflict country, they are also likely to have some interests that are rather less legitimate. Around the world, neighbors often have problematic relationships: after all, they are overwhelmingly the main source of external threat. Pakistan, which as I write is imploding following the death of Benazir Bhutto, is not going to share its sovereignty with India; Eritrea is not going to share its sovereignty with Ethiopia. So neighborhood-shared sovereignty is not going to work. The African Union recognized this when it proposed that the African peacekeeping force for Somalia should be composed of forces from any willing country *other than a neighbor*. However, Somalia also demonstrated the limits of that approach: the only country with a sufficiently strong interest to send a major force was neighboring Ethiopia.

Problem number two: The neighbors of a country that becomes post-conflict are not a natural political grouping. As a result they have no experience of cooperating as a group. Worse, their cooperation would clearly be time-limited: it may last for only a decade. Worse still, there may be rather a lot of neighbors. Take the Democratic Republic of the Congo, which is currently a post-conflict country, and look at a map. What grouping of neighbors do we get: Congo-Brazzaville, the Central African Republic, Sudan, Uganda, Rwanda, Tanzania, Zambia, and Angola. Experimental games find that cooperation gets harder as the number of players is increased, and eight participants is a lot. They also find that the players gradually learn how to cooperate, so that starting up as a group from cold would impose a phase of mistakes just when the post-conflict country is most vulnerable. Finally, one of the most basic results of experimental games is that players evolve a tit-for-tat strategy that enforces cooperation: players avoid unreasonable decisions because they would eventually get their comeuppance. So temporary cooperation is much harder than permanent cooperation.

These problems persuaded me that sharing sovereignty with the neighbors is out. What then is the alternative? The solution, I think, is to place the *legitimate* interests of neighbors in trust with a more permanent grouping that does not itself have strong interests. While this might be a regional body such as the African Union, the more obvious locus is the United Nations, and more specifically the Peace-Building Commission, which was established in 2005 and is jointly under the Security Council and the General Assembly. So, in the case of the Democratic Republic of the Congo, the United Nations would hold a share of sovereignty on behalf of the neighbors, tasked with minimizing the shared costs inflicted on the neighbors. To be clear about this, the United Nations would not, *in its own right*, hold a share of sovereignty. This is far short of the old model of United Nations trusteeship that some scholars have suggested should be revived. The post-conflict government would share sovereignty rather than be stripped of it, and the objectives of the regional or international body with which sovereignty was shared would be predefined to be the protection of the legitimate interests of neighbors.

What would guide the decisions of the trustees? To an extent, each decision has to be based on the totality of the circumstances that make each decision unique. But it would help to have an explicit set of guidelines. Guidelines are useful where different players have to be coordinated. I think that there are three distinct players. Some governments should provide or finance peacekeepers; some governments should provide aid; and the post-conflict government should reform economic policy, cut its military spending, and, if it chooses to hold elections, let them be free and fair. The embargoes on imports of armaments that are appropriate in post-conflict situations have routinely been broken by rogue companies in countries that have to date been below the radar screen of international scrutiny, but we now have the means to identify such breaches.

Each of these depends upon the others. Peacekeepers are less likely to get killed where arms embargoes are effective. The realistic exit strategy for peacekeepers is economic development. In turn, economic development is enhanced by aid and policy reform. Elections as commonly practiced to date, which is to say far from free and fair, have increased the risk of violence, not reduced it. Not only are these decisions interdependent, but they need to be sustained for around a decade, whereas to date all three players usually focus only on the short term. Guidelines that set out the mutual responsibilities of all players over the course of the decade could not be legally binding, but they could create a common expectation of behavior. They are also very much in the spirit of modern international cooperation: from the Monterrey Consensus to the United Nations Global Compact with large corporations, the approach has been to spell out mutual responsibilities. As the inclusion of the responsibility to comply with arms embargoes demonstrates, the responsibilities extend broadly and are not polarized between the governments of the rich world and those of the bottom billion.

In setting out guidelines for the behavior of each party, a post-conflict compact would also, implicitly or explicitly, reveal the red lines that should not be crossed. The clearer the red lines, the less likely they would be breached, and so international involvement in post-conflict situations would become less of a nightmare. If some red lines had been in place, I think that the post-conflict nightmares might have been avoided.

I HAVE PUT FORWARD THREE proposals for international action. Between them they address the abuse of democracy in the acquisition of power, the misuse of power once acquired, and the structural insecurity that has beset the societies of the bottom billion. Might they be adopted?

At present, discussion of international action ranges across the extremes. Think of the different stances on Zimbabwe. At one extreme there have been calls both from within Zimbabwe, by the Archbishop of Bulawayo, and from a range of international commentators, for international military intervention to depose President Mugabe. Tony Blair vetoed Mugabe's attendance at the Commonwealth Conference, and Gordon Brown refused to attend the Africa-Europe Summit because Mugabe was included. At the other extreme has been the indignant solidarity of African presidents, manifested in the election of Zimbabwe as chair of the United Nations Human Rights Committee. The three proposals in this book are a very long way from using military intervention for regime change. I think that externally imposed regime change tramples on the unhealed wound of colonialism and so is unrealistic. They are also far from noninterference. In an interconnected world, untrammeled national sovereignty leads unswervingly to hell. The proposals are a compromise between positions that are currently deadlocked.

If they were adopted, would they make a difference?

The disaster unfolding before my eyes as I finish this book is Kenya. As the book has built up I have tried to show how the rule to bolster the conduct of elections would quite probably have changed the course of Kenyan history. But the disaster that has overshadowed Africa for the past decade has been Zimbabwe. Manifestly, President Mugabe has systematically dismantled both the democratic polity and the economy of his country. So what might have averted this disaster? The only power that might realistically have changed the course of Zimbabwean history is its own military. The African Union now has a rule refusing to accept coups as legitimate. While it is entirely understandable that incumbent presidents would happily agree to such a rule, it is misplaced. Zimbabwe needed a coup, but not one that led, as in Cote d'Ivoire and Ethiopia, to further ruin. Coups need to be harnessed, not eliminated: the core proposal of this book.

Chapter 10

ON CHANGING REALITY

W E HAVE COME FULL COURSE. The societies of the bottom billion are *structurally* insecure and *structurally* unaccountable. Despite recent years being the most successful period of economic global growth on record, the appalling consequences are apparent to all. Structural insecurity hit the headlines in 2007 first due to Somalia and then to Sudan. Structural lack of accountability in the conduct of elections hit the headlines in 2007, first in Nigeria, then in Pakistan, shortly followed by Kenya. The year 2008 started with a rebellion in Chad and a coup attempt in East Timor, the president of which is currently recuperating in Australia. I fear that there will be many more such events.

So what, in a nutshell, is the structural problem faced by the countries of the bottom billion? It is that *they are too large to be nations yet too small to be states*. Too large, because they lack the cohesion needed for collective action. Too small, because they lack the scale needed to produce public goods efficiently. Societies can function well enough without some public goods because they can also be supplied privately. Some of the health and education services that are supplied as public goods in Europe are supplied as private goods in America. But some other public goods cannot be substituted by private activity. Security and accountability are such goods.

No society can rely successfully upon private security, although periodically it has been tried. The private forces hired for defense become predatory on the very people they are supposed to protect. Britons tried it after the Romans departed, hiring a gang of Jute thugs to defend them against the Picts. It took the thugs fifteen years to work out the obvious: they then slaughtered the British political elite and took over. As for the private provision of accountability, most instances that appear to be private provision, such as the way that the American health system is disciplined by the fear of being sued, depend upon being backstopped by the rule of law. In the absence of the rule of law the need to maintain a good reputation within a small network of associates can enforce a degree of accountability. Economists regularly parade the example of how thirteenth-century Jewish traders conducted long-distance trade despite the lack of law. But Avinash Dixit has recently shown that if such networks are scaled up the whole system crashes. Security and accountability are either provided by government or they are not provided. Their absence produces socioeconomic conditions such as the bottom billion have lived through for forty years. During that time they have become the poorest people on earth.

With sufficiently visionary political leadership, the states of the bottom billion could build a shared identity within the society, thereby transforming state into nation, and cooperate with the other nations of their region. Combined, these approaches would enhance the supply of the public goods, providing the security and the checks and balances that their citizens need. From time to time people capable of such leadership gain political power, but not very often. It is not by chance that the visionary leaders Julius Nyerere, Sukarno, and Nelson Mandela were all *founding* presidents. Once political power can readily be won by the self-serving, the self-serving will step forward to try their luck and the honorable will step back. Bad currency drives out good. In this book I have spared you the fancy terminology of economics, but since you have reached the end you can take delight in one technical term: in economic language the quality of political leadership is endogenous. As a result, in these societies visionary leadership is now rare.

There is thus a powerful case for security and accountability to be regarded as basic social needs that, as a default option, should be provided internationally. After the intervention in Iraq, many people might reasonably feel that the unintended consequences of security interventions are such that intervention in any form is too risky. Yet international military intervention has had many successes. The lesson is not that it is intrinsically risky, but that the circumstances that warrant it should be limited and clearly delineated.

Since any such proposal will be met by a chorus of outrage from the beneficiaries of presidential sovereignty, the five billion who live in territories that have been more fortunate can readily justify their natural proclivity to stand back and watch. It will also be the conclusion of those in thrall to a sense of victimhood: the rest of the world has already done enough damage. My own attitude used to be "just give it time." After all, in the countries that are now developed, the transition from the effective but unaccountable state of the nineteenth century to legitimate and accountable democracy took decades. But I now think that far from being on a steady progression from political violence to accountable and legitimate democracy, the bottom billion have headed into a cul-de-sac: competitive elections without restraints will frustrate internal cooperation, and presidential sovereignty will frustrate external cooperation.

This book has proposed a way to break this impasse. With minimal international action it should be possible to harness the potent force of domestic political violence for good instead of harm, thereby supplying the missing public goods. Some of the public goods will directly meet material needs: goods such as the electricity and international transport routes that have been so chronically undersupplied because of the prolonged failures of collective action. This is the role of aid as conventionally envisaged. But the key missing public goods require new instruments. International peacekeeping and over-the-horizon guarantees are politically difficult, but they work. Although expensive, they are cost-effective. International rules and standards, some voluntary, others enforced by incentives, are neither politically very difficult nor expensive. They have no significant downside, and so we should explore their potential.

The last time a secure zone of prosperity really got serious about the insecurity of a region that could not rely upon its own efforts was in the late 1940s. The zone of prosperity was America and the region of insecurity was Europe. America was motivated by both charitable concern and enlightened self-interest. Whatever the motivation, it knew it had to get serious.

What did America do? First and foremost, it transformed its security policy. The prewar strategy of isolationism was torn up: America created NATO, the system of mutual security guarantees, and placed more than one hundred thousand troops in Europe for more than forty years. America also transformed its policy toward international rules and standards. Whereas after the First World War it had treated national sovereignty as if it were an eleventh commandment, refusing even to participate in the League of Nations, after the Second it established the United Nations, the International Monetary Fund, and the Organization for Economic Cooperation and Development, and encouraged the formation of the European Community. And, yes, America found the money to help post-conflict reconstruction. It launched the Marshall Plan, and it founded the International Bank for Reconstruction, the afterthought for which—"and Development"—now gives the rebranded World Bank its important role. For good measure, America even tore up its protectionist trade policy, but that is another story. You get the picture: faced with a security danger America got serious; no viable strategy was neglected. It worked: the threat from the Soviet Union is over, but even with this massive response it took more than forty years of sustained effort.

Is the challenge facing our generation greater or less than that? The zone of prosperity has expanded enormously: the burden can now be widely shared. The region of insecurity has not shrunk, it has moved: in 1945 the societies of the bottom billion were secure because they were part of empires, now they are insecure because they are on their own. The danger is also less stark: the Democratic Republic of the Congo is not pointing missiles at Washington. In fact, we are back to 1919: it is because the dangers are amorphous that we have not got around to facing them. The failure to get serious at the Paris Peace Conference in 1919 took twenty years before the catastrophic consequences became unmistakable.

The failure to get serious since the end of the Cold War is manifested in wild swings in strategy. Sometimes we try total neglect: we left Somalia without a government for more than a decade in the hopes that it would sort itself out without us. Al Qaida eventually occupied the resulting vacuum. Iraq 2 is at the opposite end of the spectrum: preemptive total intervention. I doubt whether there is now much appetite to make that a normal part of our strategy, we're more likely to swing back to total neglect. Yet the lesson of how America overcame the threat from the Soviet Union is that faced with challenges of this scale, we will need to apply a consistent set of policies for a long time. Of course, the rationale for doing something extends beyond our own security. A billion people are living pitifully while the rest of us have credible hope of the good life. That is not just a looming security nightmare, it is a present scandal.

But self-interest and compassion are not rivals: they can coalesce into a sense of common purpose. The political right needs to recognize that its well-founded security fears should empower a more effective strategy than Iraq 2. The political left needs to recognize that guilt-ridden inaction in the face of political violence is an evasion of responsibility. The powerful emotions of fear and guilt have fogged our thinking. In the alliance of compassion and self-interest, compassion will provide the energy to get started, and self-interest will ensure that we stay the course. President Bush was right that prevention is often going to be the right response to these security problems, but he was wrong to think that the best preemptive policy was military invasion. We have a whole armory of policies at our disposal. Some of them take time to work: think decades rather than weeks. But the good news is that we are facing this problem only because we have been so incompetent at dealing with it. Had we woken up to the new problem at the time we were freed from the burden of the Cold War, we would be well on the way to fixing it. Instead we were naïve and we were selfish. It is time to put it right.

ACKNOWLEDGMENTS

THE IDEAS IN THIS BOOK are all founded on statistical research. That does not make them right, but it does make it possible to know approximately how much confidence can be placed in them. In part, that comes with the statistics, but it also comes because my work is produced within the modern academic community. The modern academic community is to an idealized community what *The Simpsons* is to an idealized family. Essentially, academics fight a zero-sum game over reputation in which the fast route to success is to demolish some prominent piece of work. You can rest assured that droves of academics on the make are hacking away at the propositions in this book. And, of course, being scared to death of them, I have done my best to protect myself by eliminating the errors. This, incidentally, is why you should be wary of all those seductive ideas peddled by heterodox thinkers. Because they are not taken seriously by the academic community, there are no kudos in demolishing them.

I greatly admire the lone academic geniuses, but I find that I work much better in a team. I depend upon a bunch of young researchers with skills far beyond my own. Most of the research on which this book is based has been done with them. With Anke Hoeffler I worked on the causes of civil war, on arms races, and on what makes a country prone to coups d'état—potentially the most sellable work I have ever done, since it is the key fear of presidents in the countries I visit. Our work on coups ended up as a different kind of race: we managed to finish it in the days before Anke gave birth to her first child. I promptly found myself in the same race with Lisa Chauvet, with whom I have worked on elections, on the costs of failing states, and on why reform is so slow. With my female workforce on maternity leave, much of the work on which this book is based has been done with young men. Both Dominic Rohner and Benedikt Goderis left Cambridge to come and work with me. With Dominic I did the disturbing work on political violence in low-income democracies that underpins chapter 1. The work with Benedikt proved so astonishing that it will form my next book: that is why neither the commodity booms nor the impact of China feature here. Along the way I worked with Mans Söderbom on how to reduce the risk of going back to violence in post-conflict conditions and with Chris Adam and Victor Davies on the role of aid in post-conflict stabilization.

Surely the most extraordinary research in this book is that with Pedro Vicente. Randomized experiments are currently all the rage in economics, but I think we are the first to have done one on how to curtail the violent intimidation of voters by corrupt politicians. Evidently, if that is what you are going to study, it's not much use choosing a parish council election in Switzerland. The setting for our research was the presidential election in Nigeria. A presidential election in Nigeria is not a tea party, as someone nearly said.

With Havard Hegre I estimated the costs and benefits of strategies to curtail violence in post-conflict situations. Cost-benefit analysis is completely standard as a tool in policymaking: a road planner would use it in deciding whether to build an overpass. But applying the technique to whether United Nations peacekeeping is good value is a stretch. At least, however, with such a cost-benefit analysis all the steps are transparent: other researchers can challenge, improve, or mock. While policymakers cannot be expected to base their peacekeeping decisions exclusively on such analysis, it does serve as a counterbalance to the other ingredients into the decision process, wise, shrewd, and politically neutral as they doubtless are.

All our papers can be downloaded from my Web site: most are also published in academic journals. Together with the articles by other scholars on which I have drawn, they are the foundations for this book. I am afraid that some are not an easy read. They carry the turgid baggage of modern scholarship. In this book I have set all that aside and tried to write something that you can enjoy. But you can read this book with both the confidence that it is well founded (though not necessarily right), and the excitement of new discovery: racing through it will take you to the frontiers of my knowledge as surely as if you were plodding through one of the underlying articles.

I have enormously benefited from discussions with three intellectual heavyweights: Robert Bates, Tim Besley, and Tony Venables. Quite possibly after reading this book they will wish that I had benefited even more: discussion does not imply agreement. Finally, I acknowledge my greatest debt, to my wife, Pauline. Not only has she been my life support system, her own experience of the societies I analyze is at least as deep as my own. Her gentle but severe comments on the first draft of my previous book, *The Bottom Billion,* spurred me into a desperate attempt to salvage something from the impending ruin of my reputation. It seemed to work, and I hope she has done it again.

APPENDIX: THE BOTTOM BILLIN

T HE COUNTRIES OF THE BOTTOM billion are defined as low-income countries that were caught in one or other of four development traps. The traps are explained in *The Bottom Billion*. This list was measured on data for around the millennium. I was reluctant to publish it for fear of typecasting: the traps are not iron laws, and a few of these countries may already have broken free. However, a list helps to focus international effort.

Afghanistan	Chad
Angola	Comoros
Azerbaijan	Congo, Dem. Rep.
Benin	Congo, Rep.
Bhutan	Cote d'Ivoire
Bolivia	Djibouti
Burkina Faso	Equatorial Guinea
Burundi	Eritrea
Cambodia	Ethiopia
Cameroon	Gambia
Central African Republic	Ghana

Guinea

Guinea-Bissau

Guyana

Haiti

Kazakhstan

Kenya

Korea, Dem. Rep.

Kyrgyz Republic

Lao PDR

Lesotho

Liberia

Madagascar

Malawi

Mali

Mauritania

Moldova

Mongolia

Mozambique

Myanmar

Nepal

Niger

Nigeria

Rwanda

Senegal

Sierra Leone

Somalia

Sudan

Tajikistan

Tanzania

Togo

Turkmenistan

Uganda

Uzbekistan

Yemen

Zambia

Zimbabwe

RESEARCH ON WHICH THIS BOOK IS BASED

T HIS BOOK IS BASED PARTLY on my own research and partly on that of other scholars. My research is posted on my Web site, which can be reached by typing my name into Google. The main published research underlying the book is:

By the author
"Post-Conflict Reconstruction: What Policies Are Distinctive." *Journal of African Economies* (forthcoming).

"International Political Economy: Some African Applications." *Journal of African Economies* 17 (2008): 110–139.

"Implications of Ethnic Diversity." *Economic Policy* 16, no. 32 (2001): 127–166.

With Anke Hoeffler
"Unintended Consequences: Does Aid Promote Arms Races?" *Oxford Bulletin of Economics and Statistics* 69, no. 1 (2007): 1–27.

"Civil War." In *Handbook of Defense Economics*, vol. 2, edited by Keith Hartley and Todd Sandler, 711–739. Amsterdam: Elsevier, 2007.

"Military Expenditure in Post-Conflict Societies." *Economics of Governance* 7 (2006): 89–107.

"Greed and Grievance in Civil War." *Oxford Economic Papers* 56, no. 4 (2004): 563–595.

With Anke Hoeffler and Dominic Rohner
"Beyond Greed and Grievance: Feasibility and Civil War." *Oxford Economic Papers* (forthcoming).

With Anke Hoeffler and Mans Söderbom
"Post-Conflict Risks." *Journal of Peace Research* (2008).

With Robert Bates, Anke Hoeffler, and Steve O'Connell
"Endogenizing Syndromes." In *The Political Economy of Economic Growth in Africa, 1960–2000*, edited by Benno Ndulu, Steve O'Connell, Robert Bates, Paul Collier, and Chukwuma Soludo, Cambridge: Cambridge University Press, 2008, 391–418.

With Dominic Rohner
"Democracy, Development and Conflict." *Journal of the European Economic Association* 6, nos. 2–3 (2008): 531–540.

With Christopher Adam and Victor Davies
"Post-Conflict Monetary Reconstruction." *World Bank Economic Review* 22 (2008): 87–112.

With Lisa Chauvet
"What Are the Preconditions for Policy Turnarounds in Failing States?" *Conflict Management and Peace Science* (2008).

With Lisa Chauvet and Havard Hegre
"The Security Challenge in Conflict-Prone Countries." In *Copenhagen Consensus*, 2nd edition, edited by B. Lomberg. Cambridge: Cambridge University Press, 2008.

By other scholars

Alberto Alesina and Eliana La Ferrara. "Ethnic Diversity and Economic Performance." *Journal of Economic Literature* 43, no. 3 (2005): 762–800.

Abigail Barr. "Trust and Expected Trustworthiness: Experimental Evidence from Zimbabwean Villages." *Economic Journal* 113 (2003): 614–630.

Tim Besley. *Principled Agents? The Political Economy of Good Government.* Oxford: Oxford University Press, 2006.

Tim Besley and Masayuki Kudamatsu. "Making Autocracy Work." *CEPR Discussion Papers*, no. 6371 (2007).

Stefano DellaVigna and Eliana La Ferrara. "Detecting Illegal Arms Trade." *NBER Working Papers* no. 13355 (2007).

Avinash Dixit. *Lawlessness and Economics: Alternative Modes of Governance.* Princeton: Princeton University Press, 2004.

Azar Gat. *War in Human Civilization.* Oxford: Oxford University Press, 2006.

Patrick Geary. *The Myth of Nations: The Medieval Origins of Europe.* Princeton: Princeton University Press, 2002.

Philip Killicoat. "Cheap Guns, More War? The Economics of Small Arms." M.Phil. economics thesis, University of Oxford, 2006.

Mwangi Kimenyi and Njuguna Ndung'u. "Sporadic Ethnic Violence: Why Has Kenya Not Experienced a Full-Blown Civil War?" In *Understanding Civil War (Volume 1: Africa)*, edited by Paul Collier and Nicholas Sambanis. Washington, D.C.: World Bank, 2005, 123–156.

Edward Miguel. "Tribe or Nation? Nation-Building and Public Goods in Kenya Versus Tanzania." *World Politics* 56, no. 3 (2004): 327–362.

Colin Renfrew. *Prehistory: The Making of the Human Mind.* London: Weidenfeld and Nicolson, 2007.

INDEX